A New Year's Present from a Mathematician

A New Year's Present from a Mathematician

By
Snezana Lawrence

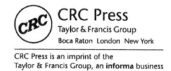

CRC Press
Taylor & Francis Group
Boca Raton London New York

CRC Press is an imprint of the
Taylor & Francis Group, an **informa** business

CRC Press
Taylor & Francis Group
6000 Broken Sound Parkway NW, Suite 300
Boca Raton, FL 33487-2742

© 2020 by Taylor & Francis Group, LLC
CRC Press is an imprint of Taylor & Francis Group, an Informa business

No claim to original U.S. Government works

Printed on acid-free paper

International Standard Book Number-13: 978-0-367-21936-9 (Paperback)
International Standard Book Number-13: 978-0-367-21937-6 (Hardback)

Visit the Taylor & Francis Web site at
www.taylorandfrancis.com

and the CRC Press Web site at
www.crcpress.com

To my Kit, who is constantly mathematizing

Contents

List of Figures

Preface

W HAT KIND OF A present do you give to a person who has almost everything? What is the best present you can ever imagine receiving? Of course, there is health, happiness, love, security, but when that is not on offer, what can possibly be given to a friend who has almost everything? These kinds of thoughts occupied one of the mathematicians featured in this book whilst he walked around a snow-covered city in central Europe, hundreds of years ago. And the present he came up with? A mathematical exploration and discovery: a present to last an eternity, and a gift to his friend and to the rest of humanity at the same time.

Is this the type of thing that mathematicians do? To answer this question, I will here describe twelve scenes to demonstrate how mathematics happens and who are the people that are engaged in it, from antiquity to the twentieth century, all seen from our own perspective – from our own place and time. I will try to explain what the diagrams, people, and objects in such scenes signify, and what the mathematicians in question did, how they did it, and, more importantly, why. The reason I chose those particular scenes are explained in each individual chapter, but the general idea is that I have chosen them and not some other moment in their lives as they signify something particularly important and memorable: a new discovery, an event that defined their work and often their lives, the friendships they made and through which they gave to the world their contribution to be used in perpetuity. These scenes will not necessarily give you the most important contributions of those mathematicians, but they will bring us closer to them as people and to their work as part of the great universal mathematical project.

This book is written in twelve chapters, with each chapter assigned to a month of the year that links to a famous mathematician or, for example, to the building they helped construct or to an organization to which they belonged. Through communicating and working with mathematicians, I found that they are as ordinary and as extraordinary as you may imagine.

However, there are some things that mathematicians do differently from others, and that is what I will focus on and share with you over these twelve chapters.

The most intriguing aspects of mathematics are those related to the reality of mathematical inventions – some are amusing, some almost unbelievable, and some will just take you to another topic, so you can keep going and learning more about mathematics. How does a mathematical discovery happen? What is it that mathematicians do that can end as an invention? And who gets to be called a 'mathematician'? These are the questions this book is dedicated to. You may know a lot about mathematics already, but I hope that you will still be able to know more after you've read this book. Or you may know little or almost no mathematics, yet I am sure that you too will be able to learn from this book where to delve deeper into mathematics, and what to pick from the vast archive of abstract thought that mathematics has ways of neatly organizing.

Mathematics, to me and to many a mathematician, is like an art, a poetry of the mind. It is a language through which we attempt to understand how things work – in this world and in any other imagined one. This book is a little homage to this particular view of mathematics and an attempt to read to you some of its nicest poems. Or better still, imagine a slow song being played in the background of our mathematical scenes!

These songs are played all the time. There is always a mathematician somewhere thinking about some mathematical thing on their own or with their friends and inventing new mathematics. How this is done and why it is important, and beautiful, and how it changes the rest of us and the world we live in, is best told through examples. So, let's go on our little tour of some of the most beautiful scenes from mathematical history. There will be a scene to read about, absorb, and imagine for each month of the year. Let us begin our journey.

Acknowledgments

S OME OF THE STORIES in this book have been used in the many articles I have written over the past decade for *Mathematics Today*. This
bi-monthly journal is a publication of the Institute of Mathematics and
its Applications, the largest professional association of mathematicians in
the UK.

It is one of the most open-minded and inclusive groups of people I have
ever come across, and this book is a celebration of that spirit that in some
way also embodies the universality of any mathematical association: where
gender, politics, colour of the skin, or religious affiliation matter not, and
the love of truth and the journey towards it are all important.

About the Author

Snezana Lawrence is a mathematical historian, with a particular interest in the links between mathematics, architecture, and the belief systems related to mathematics. Her work on the creativity, identity, and engagement in the learning of mathematics has taken her to be involved in national and international initiatives to promote the use of the history of mathematics in mathematics education.

Twitter
@snezanalawrence
@mathshistory
www.mathsisgoodforyou.com

Introduction

TWELVE DROPS OF MATHEMATICAL WISDOM

*Here we begin our journey through the large mathematical land-
scape that we represent by a desert. Not an unpleasant one but a
most beautiful desert, in which it could be nonetheless (metaphor-
ically) deadly to get lost in. We will navigate carefully, seek visible
paths and hope we will come to oasis after oasis on our journey. We
must also look after our supplies as they need to see us through the
next twelve months of the year. In this Introduction we will choose
the starting and the ending points, and enumerate our most precious
supplies of all – the precious drops of mathematical wisdom to help
us on our journey through the desert.*

WHERE SHALL WE BEGIN our journey from? What time and place in
the past will we take as the starting point and how will we navigate through these twelve chapters? I suggest that the great library of
Alexandria – one of the earliest great libraries, at the delta of the River Nile
in Egypt – could be a good place from which we can begin our metaphor-
ical journey.

The city of Alexandria was founded in the fourth century BCE by
Alexander the Great (356-323 BCE) and within a hundred years became a
cosmopolitan center of learning and culture. Around the time of the dyn-
asty of Egypt's ruler Ptolemy II in the third century BCE the great library
was established and quickly attracted some of the most important minds of
the Mediterranean at the time, amassing a huge collection of books. These

books – scrolls of papyrus or parchment, sometimes rolled and sometimes folded like accordions – contained the most important works of Hellenic culture. You could find here diverse works from contemporary poetry to mathematical books.[1]

What type of activity was going on in this old library? We can only tell by the books that have survived to this day, in addition to the records of some of the people who worked there. It was here that we would be able to see and find Euclid's *Elements*, probably the most popular mathematical collections of books ever to have existed, or, if we were to come at the right time we could see Eratosthenes of Cyrene (276–194 BCE), the librarian of this great institution. He went around taking notes of the content of the library to make its catalogue, and would be inspired by the number of things he found lying around and every so often think that something he came across was really special. As special as a prime number perhaps? Eratosthenes invented something we still use – a prime number sieve. He also calculated in his spare time the circumference of the Earth.

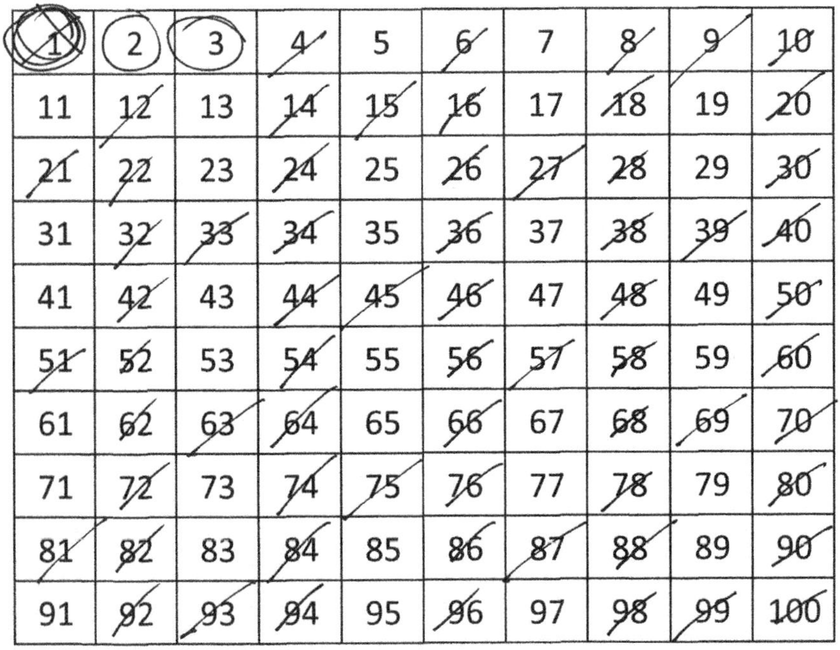

1	2	3	4	5	6	7	8	9	10
11	12	13	14	15	16	17	18	19	20
21	22	23	24	25	26	27	28	29	30
31	32	33	34	35	36	37	38	39	40
41	42	43	44	45	46	47	48	49	50
51	52	53	54	55	56	57	58	59	60
61	62	63	64	65	66	67	68	69	70
71	72	73	74	75	76	77	78	79	80
81	82	83	84	85	86	87	88	89	90
91	92	93	94	95	96	97	98	99	100

FIGURE I.1   Illustration of Eratosthenes Prime Number Sieve, which you can complete as follows: circle the next number (in this case 5) and then cross all of its multiples. Continue until you exhaust your number sieve. The numbers that are circled are primes, and crossed over ones are not (they are multiples of primes).

At the zenith of this great library's life, it is believed that it held something between forty and four hundred thousand volumes. We know that the library eventually ran out of shelf space, sometime just after Eratosthenes was succeeded by the fourth librarian Aristophanes of Byzantium (257-180 BCE), as at that time a new storage center was built in a suburb of Alexandria overlooking the sea. The library was built in Serapeum, which according to a legend, was one of the most magnificent of all temples of Alexandria, housing at the same time a temple and a library collection.

We will not hang around to see how the decline of the library began, or how, when a government of the time began to interfere with the scholarship the scholars dispersed and left, the learning declined and then of course by accident or design a number of unhappy events took place, like the fire which destroyed much of the collection during Ceasar's Civil War in 48AD. Let's just remember that nothing is permanent and that every great undertaking must be finished and a new one started somewhere else. And that there are precious moments in which great learning takes place and great books are collected and written, and these moments need to be remembered so that they could be recreated ever so slightly differently in other times and places.

This great library, situated on the shore of the Medditeraenan Sea is the starting point for our caravan to venture into the desert of the unknown mathematical landscape that will be revealed to us as we look at the scenes of mathematical activity for each month of the forthcoming year. Like all desert caravans, we will need some precious water. By a stroke of the keyboard on my computer, I've conjured into being this imaginary water to be given to us in twelve drops of mathematical wisdom over the impending twelve chapters. This should keep us fresh and healthy just as water would do on a journey through the desert.

Of course, we must not get lost in this vast and beautiful desert where mathematics can be imagined: We must always remember that the past is never only just something that has happened before us, but is actually different from that which is happening now.[2] This means that, although mathematics is universal, past mathematics is also different in some ways to that of the present. And mathematicians from the past did things differently from the mathematicians of the present: Imagine how rather strange it would be to see an original portrait of Newton if he was dressed in jeans and a t-shirt! It would be equally strange to see a modern mathematician do some of the things that ancient or a historical mathematician did. This can be even stranger if they were dressed as an ancient mathematician (but don't tell that to some of my colleagues who like to dress as past mathematicians).

LET'S CREATE OUR MAP

So we start from Alexandria. This is the only place that we will stop outside of Europe as our journey is mostly confined to this continent. It is a continent that is still in flux and always seeking to find its place in the world. So perhaps it is timely that we take this journey in Europe at the dawn of the third decade of the second millennium.

From Alexandria, the first city, the first node of our map, we travel quite a long distance to the next point of interest: a far-away place to a little village near Nottingham in England, the Robin Hood county, and a house called Woolsthorpe Manor. Woolsthorpe is where Isaac Newton was born in January 1643 (Gregorian calendar).

From Woolsthorpe we cover some thousands of miles back towards the Middle East and stop in our third place, today's Istanbul, to admire the great cathedral of Hagia Sophia, perched on the European brink of the Bosporus strait. We only have a month of February to marvel in this city and the mathematics that built Hagia Sophia's dome and then to make it back to northern Europe – and back to England and London.

In March we look at some mathematics and architecture of Sir Christopher Wren, who died in March 1723. We won't stay there long, as the heat of the Great Fire of 1666 still feels very much alive. The legend tells that Wren, the designer of the new St Paul's Cathedral, came across one of the stones from the old cathedral that was burned down in the same place, marked with the word 'resurgam' – meaning 'I shall rise again'. He had this as a symbol of the fire and the resurgence of new London built after the fire and had the word inscribed under the carved phoenix at the portico of St Paul's south door. And with this picture we move onto the next point of interest, the fifth place in our journey, the city called Erlangen in Germany.

Erlangen means 'gain' in English – and there we learn of what humanity has gained in the birth of Emmy Noether. Often called the mother of modern algebra, Noether was born in this town in 1882 and died in April of 1935 in Bryn Mawr in Philadelphia in the US. Bryn Mawr is not really a town – it is a designated area between several townships, an abstract structure, just like Noether's mathematics. And after we try to describe the beauty of the things she created, we need to move to our next place, towards the beautiful city of Milan in Italy.

There another great mathematician was born in May of 1718 – Maria Gaetana Agnesi. Maria wrote one of the most celebrated mathematical books in Italian, yet it was not translated particularly well, a curve she called 'a turn' (la versiera) became known as 'the witch of Agnesi'.

From Milan, we turn south towards Florence, to commemorate in the early summer the death of a mathematician Luca Pacioli in June 1517. We find ourselves in the beautiful town of San Sepolcro, on the upper reaches of the River Tiber, a birth-place not only of Pacioli but also of the famous painters Raphael and Piero della Francesca. Here we need to consider some beautiful paintings, but before too long we need to move onto our next destination.

July signifies the power of the written word and is the birthday month of John Dee, the author of one of the most celebrated treatises about the need for mathematics, his *Mathematical Praeface*. We visit his old house in Mortlake, now in suburban London, but during his lifetime a country home in Surrey. Some magic makes it possible to think of Dee as both a mathematician and a secret agent – it was him after all that came up with the code-name 007!

August brings us to some dangerous bends in the road. Born in the decade before the Second World War, it is in this month when we celebrate Nicolas Bourbaki. He was born in a lecture theatre in Paris, in a street behind the Pantheon, the French temple to its most famous and distinguished French citizens. There, in École Normale in Paris, we visit the birthplace of the most famous imaginary mathematician of modern times.

September brings us back to the central Europe and to Budapest, the birthplace of one of the most famous European mathematicians, Paul Erdős. Erdős died in September of 1996, having managed to write 1475 papers with over 500 collaborators. A man of so many friends he sounds like a center of some kind of network. We'll explore that too in September.

Only one month away and already missing Paris? We'll go there again, as in October we celebrate the life of Jean-Baptiste le Rond d'Alembert who died there in 1783. A happy man, a rounded life, and an idea that made some vibrations last centuries – that's for the month of October. Before we return to London for a third time.

In November we remember that one of the greatest scientific institutions of all time, the Royal Society, was founded in London in November of 1660. Don't believe me? That's good – remember to always check it out! Or *nullius in verba* (take nobody's word for it); that's precisely Royal Society's motto.

As we approach the end of our journey, we finish in the most beautiful of cities, in snow-covered Prague, where Johannes Kepler, who was born on 27th December 1571, contemplated one night the mathematical structure of snowflakes. Why are they always hexagonal? We'll talk about that and some other things before we finish the story, and have a flying look over our map once again, seeing it from a distance as if we were flying away from it in a dream.

WHY THE DESERT?

Some of the readers may ask why have I used the desert as the starting point for our journey? Could I have used the rainforest instead? I may do one day, for another journey. But for this book, the desert has won my internal contest. Not only because it is a very special place for me, but because as a real landscape, a desert is always full of surprises and unimaginable beauty, and a type of place that never fails to create in me a sense of adventure. You need to be prepared for the desert and to be prepared for the unexpected. You need to be respectful and cautious, and above all cooperative with everyone you come across as your survival may well depend on it.

But there is another aspect of a desert that is deserving of comparison with mathematics. For one, where would be a better place to contemplate the number of grains of sand if you wish to think about large numbers, I ask you?! There is a chapter that tells that story in more detail – turn to Chapter 9 and stories related to Paul Erdős to read more about that.

And then there is of course the perception of mathematics being a kind of desert, and a cold and unfriendly desert at that. Of course, you may fear deserts a little. You need, for example, to learn how to navigate your way around them. You also need to learn a few words of the languages spoken there, so that if you do get lost you can communicate with its inhabitants. This has happened to me in real life! I can tell you it is pretty difficult to find directions in a desert when your sat-nav doesn't work and you don't know the local language. Likewise, with mathematics, you need to know a little to be able to explore its landscape, and have some helpful technology with you, in absence of a sat-nav, a map would do very nicely!

IS MATHEMATICS FOR YOU, OR ARE YOU FOR MATHEMATICS?

Kids may occasionally say that mathematics is boring and irrelevant, would you believe it![3] And adults that I came across during my journey so far would often think they are either good or bad at it. Or some, even when they are making their living from doing mathematics in some way, think that they are not really mathematicians. How so? I can't explain the kids' responses other than by blaming the adults. We surely must talk to kids themselves to see why that is and whether they know about some of the things that mathematicians do. But for adults there are many who think that they are just not worthy of mathematical arts. You may think this only refers to those who haven't got direct experience of higher mathematics, but I must say, I'm afraid that is not the case. I have had an experience and even did a few experiments with my mathematician colleagues themselves. On one such occasion I asked a room full of mathematicians to raise a hand if they were a mathematician. This may sound slightly illogical but there was a preceding discussion to this experiment that made me realise that some mathematicians don't consider themselves to be mathematicians. The answer to my question was intriguing – only three people out of approximately seventy admitted to being so. What were seventy or so non-mathematicians doing at a mathematical conference I asked them?[4]

From this came a little research project that showed the disparity of the views of mathematicians from within and outside of the profession. While pursuing that project, it became apparent to me that, as groups of professionals go, this one has been an incredibly *nice* one to work with. *Nice* in this case I use to mean unpresumptuous, unassuming, rather than, for example, full of people feeling superior to others. Or simply, I found almost no one in this group of mathematicians to have had an image of themselves of being superior to others in any way. Perhaps it was the sample that I was dealing with, but still, that is what I can report from this project.

It turned out that the status of a mathematician was so highly rated by the group of mathematicians that this was only bestowed by individuals from the group on those who changed the field in a major way or had a calling to become a professor. To aim high and to maintain such a discipline (within mathematics) was I thought, quite *nice*. This stance does not do much for the profession in terms of its prestige in the world where everyone seems only to strive to become the best and to continuously seek prestige and recognition. On the other hand, that is precisely why mathematicians are so unique. But this book has not been written because mathematicians are nice people. It has been written because *mathematics* is so nice. And it's certainly very good for you!

NOTES

1 If we were to imagine the old Alexandrian library, we could do worse than to start from the contemporary Alexandriana library, built in the same city to commemorate the ancient library. Whilst the ancient library attempted

to acquire and keep records of contemporary knowledge of the time, so the modern library has special collections of the records of the Internet at any given time, the Internet Archive's *Time Machine*, among other things.

2 This theme of the difference between the general periods of past and present, is very nicely given in Schiffman (2011).

3 In fact this was a most common description of mathematics in the national survey conducted in England with the secondary age children in 2004. See Smith (2004).

4 The whole story of this experiment, and the ways both mathematicians and non-mathematicians perceive who mathematician can be and who deserves to be called a mathemetician, is given in my paper, "What are we like…", Lawrence (2016).

CHAPTER 1

January

MATHEMATICAL TRIBES

Sir Isaac Newton, English mathematician and scientist, was born on Christmas Day according to the old Julian calendar 1642, but to the Gregorian calendar (the switch in England took place in 1752) this happy event took place on 4ᵗʰ January 1643. We will be forgiven therefore if we begin our journey with Newton in the month of January and look at some of the mathematics he formulated and inspired others to learn and do. Mathematics, as we will see, can be an inspiration for a whole tribe of mathematicians. This may be strange for some, as he is known to had been a person with very few friends. But as we will see in this chapter, Newton's mathematics was a revolution of its kind and brought him a steady following during and even more after his lifetime. His study of change began with his efforts to understand how the universe works and this is how he came up with his 'machina mundi', his concept of calculus, and his understanding of space and time.

B Y ALL ACCOUNTS NEWTON was an unhappy, or at least an unlucky child. His father died before he was born and his mother, when she remarried, left Isaac, aged three, to the care of his grandmother. The important thing for both Newton and for us is, however, that he learnt how to deal with the unhappiness that his circumstances and life threw at him. From being an unhappy child and writing in his notebook that he wished 'death and hoping it to some', Newton went a long way from his home, the

FIGURE 1.1   Sir Isaac Newton (1642 Julian or 43 Gregorian calendar–1727). Wash drawing by G. S., 1848, after J. Simon, 1723, after Sir J. Thornhill, 1710. Credit: Wellcome Collection. CC BY.

Woolsthorpe Manor near Grantham in Linconshire, England, to become the Master of the Mint and one of the most celebrated mathematicians and scientists of all time.

There is then, hope. Hope that if as a young man, you write something as silly as wishing death to some (in Newton's case his mother and stepfather), you can overcome unhappiness and become a great mathematician. If the history of Newton's life is to be taken as an example, then one thing is certain: In order to overcome difficult situations and difficult times one has to admit to them, and there are two things that Newton did with writing such things in his diary. Firstly, he wrote his thoughts to allow himself to repent by recognizing them, and secondly by writing them down, he was able to ask for forgiveness. Such method of religious and moral 'purification' included one crucial step – writing in such a way that others would not likely be able to understand what was written. Newton thus taught himself how to write shorthand, having adopted the system of short-writing that was developed earlier in the seventeenth century.[1] From his notebook, which is now held in the Fitzwilliam museum in Cambridge and hence called the *Fitzwilliam notebook*, we learn what kind of person Newton was aged nineteen, in the middle of 1662.

Before the Pentecost, or the Whitsunday (the seventh Sunday after Easter) 1662, Newton wrote about the things he wanted to repent.[2] Among other sins, he asked for forgiveness for the fact that he was 'eating an apple'. I must admit that does not sound terribly sinful to me, but then again, who can tell? After a few other reasonably minor misdemeanors such as this, the graduation of the bad feelings escalates and the resentment and even possibly hatred towards his mother and stepfather becomes apparent:

> *Refusing to go to the close at my mothers command*
> *Threatening my father and mother Smith to burne them and the*
>   *house over them*
> *Wishing death and hoping it to some…*

Although he managed to overcome such feelings at that time, this penchant for anger did indeed continue in his later life as most know, and if it was my aim to show you how he was capable of great persistence in pursuing arguments in his later life, we would not be short of material to talk about. But that is not my intention. On the contrary, it was his ability to inspire a whole group of mathematicians, philosophers, and even writers, to consider him as a great, even a supreme human being, one capable of intimate insight into the workings of the universe.

I WROTE YOU A POEM

Love manifests itself in many ways and having a poem written about you is certainly one such manifestation. But this was not a personal love, it was the love of a new philosophy, a new view of the world and of mathematics, all of which came from Newton, as if from a water-spring of wisdom during his lifetime. As poems go, the first one we will mention was a short one, a kind of an English haiku. The epitaph for Newton written on his grave in Westminster Abbey says what kind of standing he had by the end of his life:

> *Nature and nature's laws lay hid in night, God said 'Let Newton be!' and all was light.*[3]

Newton went to Cambridge when he was 18, graduated at 22 (in 1665) and became Fellow of Trinity College in 1667. Whilst this shows a little slow start perhaps for someone who would become a leading scientist of his generation (and, it may be argued of all generations), we must remember that his studies were interrupted by the outbreak of bubonic plague in 1665 and he was confined to his home for a couple of years. After the end of this period however, things speed up and we see Newton become Lucasian Professor in 1689, a Member of Parliament for Cambridge in 1689, and Master of the Mint in 1696.

There are quite a few more poems written about him, or dedicated to him, from some of the most famous of the British poets and scientists. Edmond Halley (the astronomer and mathematician, after whom the Halley's Comet was named), Blake, Shelley, Byron: they all wrote about Newton.[4] But these were all well-known, celebrated people, and so I choose to give you a poem from someone who was not so eminent. It seems to me that my chosen poem clearly captures the various other things that were important to Newton, such as his religious and philosophical, and sometimes, quite strange interests.

In celebration of Newton after his death, his friend and protégé, Desaguliers[5] wrote a poem with the title *The Newtonian System of the World, The Best Model of Government: An Allegorical Poem*. The poem is a little long to quote it here in its entirety, so let us look at a few things about it.[6] First, the poem sets the scene –things started to go wrong when man became corrupt, and it was Pythagoras who by discovering and teaching the system of the universe was able to amend these corrupt ways in antiquity. How did he do it? By reasoning, conjecturing, discovering the secrets of the universe through observation and mathematics. Music of course played a part there too, the Pythagorean theory of music that is, giving the beautiful method of how harmony is created by careful choice of proportions. In modern times likewise, despite its many ills, Desaguliers says in his poem, there is a possibility to create harmony and order. But to do that we needed Newton, who discovered the laws of order that pervade the universe and if we follow in his footsteps through the conscious efforts not only of our-selves individually but as a whole social movement, we can recreate that divine order, here on Earth. That is the basic meaning of the poem.

What would Newton have thought about that? We do actually have something that he once said and that seems to suggest that we can not only recreate this universal order, but that we, humans, in fact make it possible:

> *it seems to have been an ancient opinion that matter depends upon a Deity for its laws of motion as well as for its existence… These are passive laws and to affirm that there are no other is to speak against experience. For we find in ourselves a power of moving our bodies by our thought. Life and will are active principles by which we move our bodies, and thence arise other laws of motion unknown to us. And since all matter duly formed is attended with signs of life and all things are framed with perfect art and wisdom and nature does nothing in vain; if there be a universal life and all space be the sen-sorium of thinking being who by immediate presence perceives all things in it, as that which thinks in us, perceives their pictures in the brain; these laws of motion arising from life or will may be of uni-versal extent.[7]*

We *are* an important part of this universal machine of cosmos. We are, Newton seems to say, part of some kind of universal being. The will and the thought, is a crucial part of creating such universal order. Beautiful isn't it? And it is important to see that this is where he came to, some forty-two to forty-four years after he had recognized bad will towards his parents in his notebook.

NEWTON'S INFINITELY IMPORTANT AND INFINITELY SMALL QUANTITIES

Our image of a mathematician at work for this first chapter could certainly and easily be one that depicts a common scene from Newton's life: Despite all the fame he acquired, and all the important positions he was given, Newton can be seen on his own, doing experiments or writing, thinking, and reading mathematics, doing long calculations and sketches in one of his notebooks. Living in the semi-darkness of his study, constantly at work, mainly on his own.

There are several notebooks that Newton kept during his lifetime, including the one we just mentioned (the *Fitzwilliam Notebook*).[8] From his *Mathematical Notebook*, kept in the Cambridge University Library, we learn about the mathematical investigations he undertook around 1664 to around 1665. In this short period of time, as he escaped the plague in Cambridge and was forced to work in solitude at home, Newton first thought about the force of attraction that keeps the Moon in its orbit around the Earth. It is here that he, as he said to some people towards the end of his life, was prompted to consider this force as he saw an apple fall to the ground at his house in Woolsthorpe.[9]

But this is only the surface of all that went on with Newton during these two years. To think of this force, the force of gravity, was the beginning, but to have the tools to describe and prove it required much more. In these two years, Newton developed not only his theory of gravitation but developed a new theory of light, and pioneered a new approach to mathematics in infinitesimal calculus. The calculus is a much loved branch of mathematics, but often not explained, or explained in too complicated a manner to a non-mathematician. How would you then explain calculus? Not too easily, still if you have a little patience, it is a simple and yet amazing discovery.

Before we get there, we should make clear that the mathematics Newton wrote was not neatly written, or published, or collected. Prior to the publishing of *Principia* in 1687,[10] one of the most famous and precious books in the history of mathematics, which was edited with his friend Edmund Halley, Newton had no significant mathematical work in printed form. In *Principia*, for the first time, are stated his laws of motion, his law of universal gravitation, and the derivation of Kepler's laws of planetary motion.

How come a mathematician who had discovered so many things had only written so late in his life (when he was already forty-four years old)? Well, he did write some other papers, like *De Analisi* (On the Analysis of Equations) which he wrote in 1669 and circulated amongst scholars and mathematicians in Britain and around continental Europe. But the problems he came across in publishing this, as well as getting some negative reviews of his paper on the theory of light and colour in 1671, meant a long period of silence ensued. This silence was only external, as he didn't attempt to publish anything, but the investigations and work continued at a great pace. He continued to write yet did not seek to publish. Now, thanks to the Internet, we can see his mathematical manuscripts from anywhere in the world. A particularly interesting paper, which is avaliable in the Cambridge University Library from October 1666, is his manuscript entitled *Tract on Fluxions*.

Fluxions, Newton says in not so many words, is an infinitely small rate or proportion at which a quantity changes its size or magnitude. Why would that be important? And can we even picture that? We can describe quantity as a concept quite easily in mathematics. One way of doing that would be to use a coordinate system. The coordinate system was first described by the French philosopher and mathematician René Descartes in 1637.[11] The system was also independently invented by Pierre de Fermat, although he did not publish on the subject. In Descartes's book, his famous account on geometry *La Géométrie*, the coordinate system is used, yet it was another mathematician, Francis van Schooten[12], who showed how it could be used more extensively (see Figure 1.2). The system, although it used two linked variables, x and y, was drawn as having one axis to measure positions from, in order for a curve to be drawn.[13]

Of course, over the centuries (and earlier than that), the coordinate system was simplified and it now appears like the system in Figure 1.3. Any change in x will result in the change of y; this is determined by their relation which is given as a function of x. What Netwon became interested in was how to study that change – could we, at any point on the curve that illustrates this algebraic relationship, say whether it will go up or down next? Or whether it will at some point reach maximum or a minimum value? If you look at a curved line, a crooked line as Newton would call it, and you want to study it and see how it changes, you can do that by drawing tangents to it so you can observe in which direction this curve will go. That is what Newton found and the process to do this is, in the most simplest terms, what became known as calculus.[14]

Figure 1.4 needs a little more explanation. The sketch is generated through a strictly defined motion, which can also be studied by means of an equation in two variables. Newton's sketch (recreated in Figure 1.4) shows the curve that is called a quadratrix – the curve is generated through a strictly defined motion, which can also be studied by algebraic means. This sketch shows how Newton was interested in drawing tangents with such 'mechanical curves', and how the position of the tangent would change as the mechanical curve is being generated. It also shows his study of the old masters in mathematics. This curve is sometimes called the quadratrix of Hippias, after Hippias of Elias[15] and was studied throughout the history of mathematics. If I tell you how it works, you will understand the importance of drawing and knowing where that tangent to the curve is at any time – the main interest Newton had in the curve. This is then how the curve is generated: As the line segment QR moves uniformly from AB to CD, at the same time the ray OP rotates about point O uniformly from OA

FRANCISCI à SCHOOTEN
LEYDENSIS
In Academia Lugduno-Batava Mathefeos Profefforis,

EXERCITATIONVM
MATHEMATICARUM
LIBER PRIMUS.

CONTINENS

PROPOSITIONUM
ARITHMETICARVM
ET
GEOMETRICARVM
CENTURIAM.

LVGD. BATAV.
Ex Officina JOHANNIS ELSEVIRII,
Academiæ Typographi.
cIɔ Iɔc LVII.

FIGURE 1.2 The frontispiece of *Exercitationum mathematicarum libri* 1656–1657, by Franciscus van Schooten, where he described the coordinate system invented by Descartes.

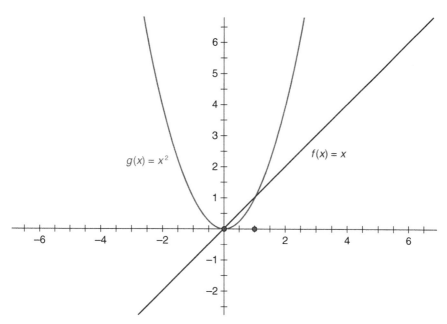

FIGURE 1.3 The coordinate system as we use it today. The function of x is shown as y, and so for example, for $y = x$ we would have a straight line, here coloured blue, or $y = x^2$, the curved (crooked line according to Newton), on our diagram coloured red. Image by author.

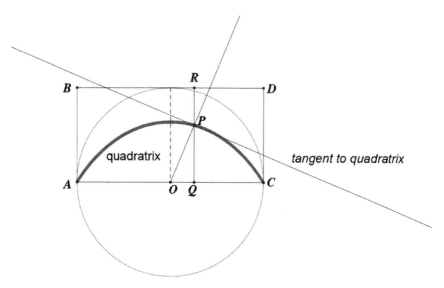

FIGURE 1.4 Newton drew a 'mechanical line' with its tangent. The drawing is reproduced by the author, showing 'mechanical line' to be quadratrix of Hippias.

to OC (clockwise in our instance, although we could do the whole process in reverse). The curve traced by the intersection of the ray OP and the segment QR is called a quadratrix.

And this is how calculus works: It is a method of studying change, the part of which is by drawing tangents to curved lines, or better, knowing at any point of the curve what the tangent would look like by its equation. Originally this part of the calculus was called the 'calculus of infinitesimals' as it dealt with these infinitely small quantities and their change. Newton had found the method through which he could study motion and the change that motion caused, and describe these mathematically. He was interested in working out how mechanical and other curves transform with a change of variables that define such curves. This was crucially important to him as he was primarily interested in the principles of change in general, and wanted to describe the phenomena in its many manifestations.

The study of change, of transformations that curves in some way signify, is what really interested Newton more than anything else: The study of how things move and what underlying force makes them behave in such ways (for example, the Moon orbiting the Earth). If seen in that light, his interest in transforming metal into gold should not seem too weird, although it led, in many ways, to some of the weirdest writings he produced.

THE SEARCH FOR THE PHILOSOPHER'S STONE

Newton was a busy man and yet he found time to write approximately a million words on alchemy. Alchemy is a type of proto-science, a tradition that originated in ancient Egypt and spread around the world. Different eras concentrated on different things and some historians of science suggest that alchemy is the precursor of chemistry. But there is something that despite the meanings it acquires in different historical periods, remains the same: The alchemist's quest to understand and use the 'transformation' as a force of nature, whatever that transformation may be. In Newton's time the alchemist's quest was to develop a method of transforming base metals, such as lead, into noble metals, such as gold. Was Newton really trying to achieve this universal goal of alchemy in his many experiments and through the reading and contemplating on the subject? I think it is even more than that: He seems to have been interested in finding ways in which to manipulate matter, as that would lead him to understanding its structure. This is such a vast subject and Newton's own writing on it is so great in volume, I cannot give it full attention here. I wanted to mention this here as it introduces an important subject that will be further explored in later chapters in this book. Newton, in his alchemical manuscripts, examines what the world is made of, and is particularly interested in the forces of nature, and their different manifestations.

In this search to discover the secrets of this material world, Newton did not only experiment and write, but read and translated. One such translation has survived: It is a translation of Tabula Smaragdina, the mythical tablet that contained the secret of the nature of matter. It is probably the closest that Newton comes to poetry himself, and it begins:

> *Tis true without lying, certain & most true.*
> *That which is below is like that which is above*
> *and that which is above is like that is below...*

The alchemy of course deals with change too, just like calculus, but using different methods. It was the part of his quest from the very beginning of his intellectual life and through his creative meddling with different aspects of understanding change, that he learnt to change himself. As he kept thinking about the mathematical, physical, theological and alchemical concepts, and worked on the writings of many of those who came before him, Newton developed new ideas. This in turn gave birth to the discovery of gravitation, his understanding of the structure of light and his discovery of calculus. Of course, one can argue, as a friend of mine once suggested, that Newton was very successful in alchemy too. He did become a Master of the Mint in his later years and as such was personally responsible for minting coins for Great Britain. He had obviously found the secret of how to transform base metals into gold after all.

THAT DIVINE CLOCK

One of the most famous stories about Newton is his animosity with Leibniz and the 'calculus' battle that raged between them. This battle turned into an all-out war, from wanting to establish and prove the supremacy of the invention of calculus, to the question about whose system of science was more accurate. The original question, was Newton first or was it Leibniz[16] who conceptualized calculus before him, does not have a simple answer: They both invented different aspects of calculus as we now know it. Quite a few books have been written about this, so I won't repeat or summarize how this happened or who did what in this battle for the supremacy of invention.

What struck me was how strange it must be, that two men, living in separate countries, interested in the study of change, using pretty similar methods in mathematics, were also both deeply committed to religious and alchemical research. Amidst so much similarity, they became arch-enemies until their very end. But how similar were they really? In some crucial aspects they represented two quite different world views. And from there on, they attracted different followers reinterpreting the natural philosophy in quite diverse ways.

Let us then see what was so different between Newton and Leibniz. Of course, I didn't give space to Leibniz so far, so you know very little about what he stood for. Nor have I given in detail Newton's philosophy or beliefs about the nature of the cosmos. Let's face it, as we will have to sooner or later, their science was one of the *aspects* of trying to work out how the cosmos works. Mathematics was part of that too – finding underlying principles and laws that can be expressed mathematically confirmed certain beliefs and gave space for their further philosophizing on the nature of the world. But they were both deeply religious, in a kind of way that mathematicians sometimes are, not completely accepting the religion of the day, but trying to work out the essence of that force which religions call 'God'. Their major disagreements around what God was like was extended to the disagreement about the structure of the universe, the concept of space and time, and the possibility of an 'action at a distance'. The latter is a particularly interesting concept that we should look at in a little more detail. To understand it, we need to look at how they achieved their different positions in regards to this 'action at a distance' and explain what it is.

Both Newton and Leibniz agreed that mathematical science of nature was crucial for the understanding of the world. They disagreed about the amount of knowledge they and, more generally, humanity could acquire: Leibniz thought that nothing can resist rational thought. At this point it may be timely to say that Leibniz was an incurable optimist, and thought that what we have is always the best possible of all possible worlds. Newton on the other hand was more modest and described his scientific method as a type of art, involving considerable rumination:

> *I do not know, what I may appear to the world; but to myself I seem to have been only like a boy playing on the sea-shore, and diverting myself in now and then finding a smoother pebble or a prettier shell than ordinary, whilst the great ocean of truth lay all undiscovered before me.*[17]

Newton also introduced a concept of an absolute space and time – whatever the framework, whatever the system we choose to believe in, one thing we can count on is the absolute space and time. This absolute container of all things means that whatever object we may observe, or how we perceive things, doesn't ultimately make a difference: Space is absolute, as is time. This was a huge thing to promote in the time in which they lived and on this one aspect Newton disagreed heartedly with Leibniz.

Leibniz thought space and time made no sense unless there is some kind of relative framework. An example of this is a rotating sphere. If there is not some kind of relative reference frame, then this sphere would not be perceived to rotate.

This was compounded with their disagreement about 'action at a distance'. Newton believed that the universe is similar to clockwork. It is perfect and set to work as a perfect machine, its movements determined by mathematics. What role does God have in it? Well, he set the thing to work, he set the clockwork in motion. God did this because the divine will to do this, to set this clockwork in motion, existed *prior* to divine intelligence. No, no, Leibniz wouldn't have it! Leibniz's view was that God only operated rationally within the laws of logic and reason.

FRIENDS IN NEED

I do admit to this being an over simplified explanation of their differences, but unless you want to spend a whole decade studying this, perhaps this is enough to get you through January. I very much think that Newton's supporters in a way helped to win the argument – at least in England, at the time. However, perhaps not in general, and perhaps this battle is in some ways still going on. Is space and is time absolute? Or are they relative depending on who perceive them? Newton was perhaps lucky enough to have lived in a society whose emerging modern networks were beginning to be formed, like those that grew in East London at the time. In 1717, the Spitalfields Mathematical Society, the first mathematical society in the world, was formed in London. There, the refugees from France like Desaguliers the Huguenot, the artisans and workers, the coffeehouse regulars, met to discuss and learn new mathematics and to better themselves. Newton's mathematics and science were here at the forefront, and Desaguliers, his faithful follower, taught his new natural philosophy there.

It is quite possible that in these settings Newton gained followers and supporters. At the time, this was a revolutionary and very modern way of doing science and mathematics and Desaguliers was very much a leading figure in this scene. As a child, Desaguliers became a refugee through the continued purge of Protestants from France in 1685.[18] His father, who was a Protestant minister, was exiled and Desaguliers traveled as a small child with his mother from La Rochelle to England. Then, some forty years later, just as Newton was dying, another French refugee came to London. He was one the greatest European philosophers and authors, François-Marie Arouet, known otherwise as Voltaire. Like Desaguliers' father, Voltaire got into trouble with the establishment in France and became a staunch critic of the Catholic Church. The luck, for us, for Newtonian science, and for some other people we will come to learn about in the May chapter, is that Voltaire made good friends with the intellectuals of London. He also met the poet Alexander Pope, whose most famous 'English haiku' we heard earlier celebrated Newton. Voltaire may have even attended Newton's funeral.

He was certainly greatly influenced by his stay in England. The hospitality and open-mindedness, the freedoms related to speech and practice of religion, the institutions and associations such as those that Desaguliers belonged to, made a great impression on him. To the world of literary culture he may be best known as the author of *Candide*, a satire, making fun in a way of the Leibnizian optimism. *Candide* is a story of a young man full of optimism, living a nice but sheltered life, an existence which changes dramatically once he encounters hardships and witnesses all the problems that life has a custom of throwing at people. Perhaps inspired by the experience Voltaire himself went through, the story goes against the motto of Candide's teacher Panglos "all is for the best" ("tout est pour le mieux") and instead argues for us to *work* in order to achieve good things in life. In contrast to the unbound optimism of Leibniz and Candide, there is on the other hand the Newtonian system of the world. In it, although everything works as clockwork, there is a place for recognition of order and for a particular way of behavior. That, as Desaguliers' poem also vividly describes, can be done by following the universal principles that Newton was able to recognize and describe in his science.

The configuration of all these aspects of disagreements between Newton and Leibniz seems to have had an effect on the great following Newton and his mathematics and science acquired after his death. Did he have more followers than Leibniz? I would not go that far, but his work certainly had an appeal for the modern man and woman.

The steady following of literati and mathematicians Newton had acquired included Maria Gaetana Agnesi (his Italian priestess and someone we will meet in Chapter 5) and Voltaire. Voltaire's friend, mistress and partner, Madame du Châtelet too wrote a book on Newtonian science.[19] How important is it to have friends in the world of mathematics? Newton did not have very many close friends during his lifetime, but his mathematics and science attracted many followers. Perhaps it was fashion, perhaps it was the degree of freedom that his philosophy entailed and gave hope to the followers, or perhaps it was that Newton's absolute space truly was more accurate a concept than the relative space of Leibniz. The current and the future explorations of the cosmos may tell us more about that.

NOTES

1 This system was invented by Thomas Shelton (1600/01–1650?). Apart from Newton, two other notable diarists used this system, Samuel Pepys, the MP and diarist of his time, most prominently of the Great Plague of London and the Great Fire of London, (1633–1703) and Thomas Jefferson (1743–1826), one of the Founding Fathers of the US, statesman, diplomat, lawyer, and architect.

2 The Whitsunday is the seventh Sunday after Easter, and traditionally a day when Christians celebrate the descent of the Holy Spirit upon Christ's disciples.

3 Newton died in 1727 and this epitaph was written by Alexander Pope (1688–1744).

4 Edmond Halley (1656–1742), English astronomer, mathematician, meteorologist and physicist, wrote a poem *For the Principia* in 1687, which was published as a Preface to *Principia's* first edition. He funded Newton's publication of *Principia* and used the laws of motion described there to compute the periodicity of Halley's Comet in 1705.

William Blake (1757–1827), English poet, painter, and printmaker, wrote two poems that are dedicated to Newton – one exclusively so, *You don't believe*, written around 1800–10, and another, his famous *Jerusalem: The Emanation of the Giant Albion*, written between 1804–20, which mentions Newton.

Percy Bysshe Shelley (1792–1822), English poet, wrote his *Queen Mab: A Philosophical Poem* in 1813 in nine parts, of which part V was dedicated to Newton.

George Gordon Byron, or better known as Lord Byron (1788–1824), father of Ada Lovelace (1815–1852), a mathematician and writer, wrote a satirical poem *Don Juan*, around 1819–24. In it Byron mentions Newton, painting a picture of him sitting under the tree.

When Newton saw an apple fall, he found
In that sight startle from his contemplation...

5 John Theophilus Desaguliers (1683–1744) was born at Là Rochelle in France, and brought to England by his parents, Huguenot refugees, in 1685. In 1709 he gained his BA at Christ Church Oxford, and became the lecturer on experimental philosophy at Hart Hall Oxford (now Hertford College) in 1710. Desaguliers was a high-ranking Freemason, and became a third Grand Master of the Grand Lodge of England (in 1719).

6 The whole poem can be however reproduced here, and goes as follows:

In Ancient Times, ere Bribery began
To taint the Heart of undesigning Man,
Ere Justice yielded to be bought and sold,
When Senators were made by Choice, not Gold,
Ere yet the Cunning were accounted Wise,
And Kings began to see with other's Eyes;
Pythagoras his Precepts did rehearse,
And taught the System of the Universe;
Altho' their Observations then were few,
Just were his Reasonings, his Conjectures true:
Men's Minds he from their Prepossessions won,
Taught that the Earth a double Course did run,
Diurnal round it self, and Annu'al round the Sun,
That the bright Globe, from his AEthereal Throne,
With Rays diffusive on the Planets shone,
And, whilst they all revolv'd, was fix'd alone.
What made the Planets in such Order move,
He said, was Harmony and mutual Love.
The Musick of his Spheres did represent
That ancient Harmony of Government:
When Kings were not ambitions yet to gain
Other's Dominions, but their own maintain;
When, to protect, they only bore the Sway,
And Love, not Fear, taught Subjects to obey.
But when the Lust of Pow'r and Gold began
With Fury, to invade the Breast of Man,
Princes grew fond of arbitrary Sway,
And to each lawless Passion giving Way,
Strove not to merit Heaven, but Earth posses'd,
And crush'd the People whom they should have bless'd.
Astronomy then took another Face,
Perplex'd with new and false Hypotheses.

Usurping Ptolemy depos'd the Sun,
And fix'd the Earth unequal to the Throne.
This Ptolemaick Scheme, his Scholars saw,
No way agreed with the Phoenomena:
But yet resolv'd that System should obtain,
Us'd all their Arts his Tenets to maintain;
A Revolution in the Earth to shun,
Immense Velocity they gave the Sun;
Then Solid Orbs with strain'd Invention found,
To shew how Planets might be carry'd round.
But when th' Observers, who the Heav'ns survey'd
Perceiv'd the Planets sometimes retrograde,
Sometimes directly mov'd with hasty Pace,
Sometimes more slow, then stopping in their Race,
For this, Expedients must invented be,
That, with it self, their System may agree,
And keep some shew of Probability.
Within their thicken'd Orbs new Orbs they made,
Each Deferent its Epicycle had,
So round the Earth the Planets still convey'd
Wheels within Wheels complex'd, they thus invovle,
And yet Appearances but falsely solve.
Like Peter's Coat, the System burthen'd grew,
Keeping old Fashions, adding still the new.
But when Philosophers explor'd the Skies,
With Galielo's new-invented Eyes;
In Mercury and Venus, then were shewn
Phases like those of the inconstant Moon:
And like black Pathces, crossing Phebus Face,
These two inferior Globes were seen to pass;
Which shew'd the right of Sol to hold the central Place.
Comets, (no longer Meteors to be fear'd
As threatning Vengeance with their Tail or Beard.)
By Telescopes, were lasting Bodies, prov'd
Like Planets, in revolving Orbits mov'd,
Whose Course destroyed the Ptolemaick World,
And all the Chrystal Orbs in ruin hurl'd;
Prov'd 'em fictitious, as in empty Space they whirl'd.
So when a Minor King the Scepter Sways,
(Some Kings, alas! are Minors all their Days)
How hard's the Task, how great must be the Pains
For envi'd Regents to direct the Reins?

While jarring Parties rend the sinking State,
Machines, by Art, must bear the tott'ring Weight;
Statesmen perplex'd, with their Invention rack't,
One Day make Edicts, and the next retract;
The Coin, to Day, shall in its Value rise,
To morrow, Money sinks and Credit dies;
One Year the Minds are rais'd by specious Schemes,
The next, are wak'd from all their golden Dreams:
And now th' Expedient is a Foreign War;
And now soft Peace can ne'er be bought too dear;
And now the Work is done by Plots and Panick Fear.
But bright Urania, heavenly Virgin, say,
How th' ancient System made again its Way,
And, that Consistency might be restor'd,
The Sun became, once more, the central Lord;
What Praises to Copernicus are due,
Who gave the Motions, and the Places, true;
But what the Causes of those Motions were,
He thought himself unable to declare.
Cartesius after, undertook in vain,
By Vortices, those Causes to explain;
With fertile Brain contriv'd, what seem'd to be
An easy, probable, Philosophy;
No comuring Terms or Geometrick Spells;
His gentle Readers might be Beaux and Belles.
In Plato's School none cou'd admitted be,
Unless instructed in Geometry;
But here it might, (nay must) aside be laid,
And Calculations that distract the Head.
Thus got his Vogue the Physical Romance,
Condemn'd in England, but believ'd in France;
For the bold Britons, who all Tyrants hate,
In Sciences as well as in the State,
Examin'd with experimental Eyes,
The Vortices of the Cartesian Skies,
Which try'd by Facts and mathematick Test,
Their inconsistent Principles confess'd,
And jarring Motions hast'ning to inactive Rest.
But Newton the unparallel'd, whose Name
No Time will wear out of the Book of Fame,
Caelestial Science has promoted more,
Than all the Sages that have shone before.

Nature compell'd, his piercing Mind, obeys,
And gladly shews him all her secret Ways;
'Gainst Mathematicks she has no Defence,
And yield t'experimental Consequence:
His tow'ring Genius, from its certain Cause,
Ev'ry Appearance, a priori draws,
And shews th' Almighty Architect's unalter'd Laws.
That Sol self-pois'd in AEther does reside,
And thence exerts his Virtue far and wide;
Like Ministers attending e'ery Glance,
Six Worlds sweep round his Throne in Mystick Dance.
He turns their Motion from its devious Course,
And bends their Orbits by Attractive Force;
His Pow'r, coerc'd by Laws, still leaves them free,
Directs but not Destroys, their Liberty;
Tho' fast and slow, yet regular they move,
(Projectile Force restrain'd by mutual Love,)
And reigning thus with limited Command,
He holds a lasting Scepter in his Hand.
By his Example, in their endless Race,
The Primaries lead their Satellites,
Who guided, not enslav'd, their Orbits run,
Attend their Chiefs, but still respect the Sun,
Salute him as they go, and his Dominion own.
Comets, with swiftness, far, at distance, fly,
To seek remoter Regions in the Sky;
But tho' from Sol, with rapid haste, they roll'd,
They move more slowly as they feel the Cold;
Languid, forlorn, and dark, their State they moan,
Despairing when in their Aphelion.
But Phoebus, soften'd by their Penitence,
On them beningly sheds his Influence,
Recalls the Wanderers, who slowly move
At first, but hasten as they feel his Love:
To him for Mercy bend, sue, and prevail;
Then Atoms crowd to furnish out thier Tail.
By Newton's help, 'tis evidently seen
Attraction governs all the World's Machine.
But now my cautious Muse consider well
How nice it is to draw the Parallel:
Nor dare the Actions of crown'd Heads to scan:
(At least within the Memory of Man)

If th' Errors of Copernicus may be Apply'd to ought within this
    Century,
When e'er the want of understanding Laws,
In Government, might some wrong Measures cause,
His Bodies rightly plac'd still rolling on,
Will represent our fix'd Succession,
To which alone th' united Britons owe,
All the sure Happiness they feel below.
Nor let the Whims of the Cartesian Scheme,
In Politicks be taken for thy Theme,
Nor say that any Prince shou'd e'er be meant,
By Phoebus, in his Vortex, indolent,
Suff'ring each Globe a Vortex of his own,
Whose jarring Motions shook their Master's Throne,
Who governing by Fear, instead of Love,
Comets, from ours, to other Systems drove.
But boldly let thy perfect Model be,
NEWTON's (the only true) Philosophy:
Now sing of Princes deeply vers'd in Laws,
And Truth will crown thee with a just Applause;
Rouse up thy Spirits, and exalt they Voice
Loud as the Shouts, that speak the People's Joys;
When M a j e s t y diffusive Rays imparts,
And kindles Zeal in all the British Hearts,
When all the Powers of the Throne we see
Exerted, to maintain our Liberty:
When Ministers within their Orbits move,
Honour thier King, and shew each other Love:
When all Distinctions cease, except it be
Who shall the most excell in Loyalty:
Comets from far, now gladly wou'd return,
And pardon'd, with more faithful Ardour burn.
ATTRACTION now in all the Realm is seen,
To bless the Reign of George and Caroline.

7 Written between 1704–6, Newton Archives, Trinity College, Cambridge.
8 You can see the digitised copies of these on the Newton Project website, at www.newtonproject.ox.ac.uk/texts/notebooks.
9 Whether this actually took place or was a romantical explanation is not entirely certain. For further detail, see Rob Iliffe (2012), p.19.
10 *Principia Mathematica* is a short title for *Philosophiae Naturalis Principia Mathematica* (in Latin) or in English *Mathematical Principles of Natural*

Philosophy. It was published in three books in Latin, first published in the summer of 1687. Second and third editions, published by Newton, came in 1713 and 1726.

11 René Descartes (1596–1650) was a French philosopher and mathematician.

12 Franciscus van Schooten (1615–1660), a Dutch mathematician.

13 In the book published in 1649 by Frans van Schooten, the coordinate system got the other axis to simplify and clarify the connection between variables x and y.

14 This part or aspect of calculus is called derivation and is the opposite of integration, which aims to study areas enclosed by curves or volume. It is incredibly useful in engineering: If we, for example, 'spin' such a curve around one of the axes, we can create a solid of revolution that we can then further study through calculus.

15 Hippias of Elis, c. 430BC, was also a philosopher, a contemporary of Socrates. He lectured on mathematics and was reputedly vain. We have this information from a secondary source – Plato himself.

16 Gottfried Wilhelm von Leibniz (1646–1716), was a German philosopher and mathematician.

17 Brewster (1855), *Memoirs*, vol II, p. 407. David Brewster was a Scottish physicist and mathematician and wrote an extensive biography of Newton based on Newton's correspondence and work in optics.

18 The Calvinist Huguenots were founded around 1550 and were linked with the Swiss Protestants. The movement was one of reform and spread quickly in Gallic lands. There were several massacres of Huguenots by the Catholics; a famous one certainly was the St Bartholomew's Day Massacre. This was followed by the Edict of Nantes in 1598 that promised religious toleration. However, King Louis XIV revoked this edict in 1685, which is when Desagulier's father fled France.

19 Of Agnesi more in Chapter 5, the chapter for May. Gabrielle Émilie Le Tonnelier de Breteuil, Madame or Marquise du Châtelet (1706–1749) was a French mathematician, natural philosopher, and author. Her translation and commentary on Newton's *Principia* was published posthumously in 1759, and is still a standard French translation.

CHAPTER 2

February

THE HOLY MONTH OF WISDOM

What is mathematics for? In February we learn about one of the most famous mathematical applications and how mathematicians have always been well placed to solve the problems of not only the everyday life, but to promote and celebrate the more elusive problems: those of glory and power – for better or for worse. Hagia Sophia, a Christian cathedral completed in Istanbul (then Constantinople) on 23rd February 532 AD, was for over a millennia the largest dome structure in the world. We look at something that made it possible – mathematical knowledge of Anthemius of Tralles, his understanding of Thales' theorem and his approximation of π. Here we link Archimedes with Anthemius' approximation of π and how this knowledge was communicated and preserved in the ancient world.

In this chapter you will learn how conics are 'made' and we of course mention briefly Apollonius of Perga so as to plant a seed for one of our threads – we'll meet him later again in July.

IF YOU WERE IN any doubt, knowing how to use mathematics can be compared with making magic. Anthemius, a mathematician from Tralles, an ancient town (now part of Aydin in Turkey) some thirteen hundred years ago used this knowledge to settle a dispute with his neighbor. From the stories passed down to us about Anthemius, this one stands out. His neighbor Zeno seems to have had a penchant for loud

FIGURE 2.1   Hagia Sophia, designed by Anthemius of Tralles (c. 474–533 or 558 AD) and Isidore of Miletus (exact dates unknown). Istanbul, *Ayasofia from the Blue Mosque*. Engraving by D. Pronti after W. Reveley. Credit: Wellcome Collection. CC BY.

entertainment, a taste not shared by Anthemius. To teach him a lesson, Anthemius prepared his trick for one such loud party. He made leather tubes, the ends of which he inserted between the joists and floorboards of Zeno's room where the entertainment was to take place. The tubes were to be inflated with steam which would have made the floorboards shake as the steam travelled through the tubes. The result was that Zeno and his guests thought there was an earthquake taking place – upon which Zeno would have come out to see what was going on and was met by lightening Anthemius made by flashing his burning mirrors towards Zeno. The first trick, the one with the steam-inflated tubes under Zeno's floorboards was of course science at its best, and the second, the burning mirrors, a trick that has much to do with mathematics. Anthemius probably learnt this one from Archimedes[1], as we have records that Archimedes used 'burning mirrors' to defend Syracuse from a siege in 213-212 BC.[2] The adjective 'burning' was acquired as such mirrors were designed to focus the rays of sun in one point whereby the force of light and heat would be concentrated and cause a spark which would then start the burning of the object on which they were focused.

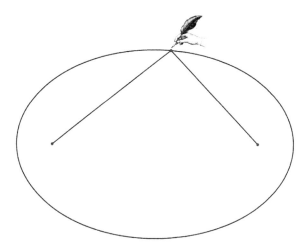

FIGURE 2.2   Construction of an ellipse by a use of a string and a pencil – first described by Anthemius of Tralles.

Whether Anthemius developed any *new* theory to construct burning mirrors is a question for a debate. But from the fragments of his treatise *On Burning Mirrors*[3] we can see that his geometrical model was based on an ellipse and that he described a construction of such an ellipse (for the first time in history as far as we can tell) by the use of a string and its foci. In order to recreate this experiement you will need a pencil, two pins or nails, and a little bit of string. For the ellipse, you need to first define its foci, which are two fixed points that will be used to construct your ellipse. Get the string around those two nails and a pencil which will stretch the string. Move the pencil and presto, you have an ellipse. Of course you can do the same by imagining the whole process.

An ellipse is a conic section that is produced when a cone is cut by an oblique plane but must not intersect the base of the cone, as in Figure 2.3. (If an oblique plane does intersect with the cone's base, we get two other types of conic sections: parabola and hyperbola.) Have you noticed something if you performed our little experiment? The length of the string remains the same although you are moving the pencil to trace the ellipse. This means that the sum of distances to the two foci from any point on the ellipse remains the same.

Although quite a clever way to construct an ellipse, this was not the most important thing that was passed to us from Anthemius. His most important contribution is his involvement in the design and construction of the magnificent church of the Holy Wisdom – Hagia Sophia – which was

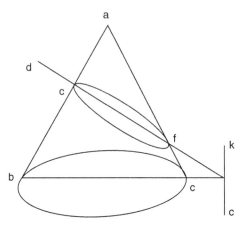

FIGURE 2.3 Conic sections are curves produced by intersection of a cone with a plane in different ways. The example given here is of a cone intercepted by a plane at an angle to the cone's axis, producing an ellipse. Apollonius, 1566.

built in Constantinople, now Istanbul, in sixth century AD, between 27th December 527 AD and 23rd February 532 AD.

This building replaced a previous church on the same site, one that had a wooden roof and was commissioned by the Emperor Constantius II around 360 AD. At the time of the first construction, the city was named after Constantius' father, Constantine I, the first Christian ruler of the Roman Empire, which then became known as the Byzantine Empire. Several iterations of the building replaced the original, but finally, during the Nika Riot in 532 AD, the church was completely destroyed by rioters. They challenged the Emperor; the crowds, until then divided into 'Greens' and 'Blues', united. It is reported that around thirty thousand people died in the riot and its aftermath. The brutality of the situation had to be somehow remedied and peace and order reestablished. The Emperor Justinian I was well aware of that. As a consequence, he set himself two goals in order to recapture the admiration and re-establish peace. His first major goal was to re-unite the Empire, reconquering the lost western half of the Roman Empire. In this he was partially successful, and is sometimes known as the 'last Roman' because of this effort. His second goal was to unite his citizens through the established religion; he therefore requested for a new church to be built on the site of the old Hagia Sophia and surpass in design and size any other existing church in the Empire. He found two mathematicians for the job, Anthemius of Tralles and Isidore of Miletus. Anthemius was known also as an architect and engineer and worked for Justinian on repairing a dam of a great fortress at Daras, the town bordering Persia.[4] Isidore of Miletus was, on the other hand, known as having worked on the mathematics of Archimedes.[5]

The two mathematicians, Isidore and Anthemius, joined forces to build the church dedicated to the Holy Wisdom (its English translation), to celebrate not only the rule of Christianity in the Roman Empire with Byzantium at its center, but also the political and military prowess of the Empire with Justianian I at its head. All cities mentioned so far – Contantinople, now Istanbul, Tralles (the town from which Anthemius came), now Aydin, and Miletus, are in modern-day Turkey. Isidore's birthplace was also a cradle of what we now call Milesian school, of which Thales was a prominent member.

GOING BACK TO THALES

We'll go back to Thales because he deserves some of the credit that we are giving to Anthemius and Isidore. Thales was the first known great Greek mathematician and lived in the sixth century BC. He came up with a method of mathematical thinking that became standard for later Greek mathematics. Thales is credited with five theorems, and one of them is named after him. But which one – there is some question over that. Is Thales' theorem the one stating that the triangle in a semicircle is a right-angled triangle (Figure 2.4), or is the theorem that deserves his name the intercept theorem: The one that says that similar triangles and ratios can be found on two parallel lines crossed by two intersecting lines (Figure 2.5)?

Whichever of these two we decide to call Thales' Theorem is really not too important, as there are three more theorems attributed to him by the late antiquity mathematical authors.[6] The intersecting lines theorem is, however, an interesting one for us, as its three-dimensional extension also appears in the study of Anthemius' burning mirrors.[7]

We therefore know that Anthemius was aware of and built on the work of Thales. This is not surprising as Euclid's *Elements* were by the time of Anthemius an ancient text, having existed for approximately eight centuries.[8] Anthemius' book on mirrors also shows how much he was interested in the application of geometry. For example, in it he discussed lifting objects more easily if they are grabbed at the center of their gravity. There he also mentions how to find a focus of a parabola in order to construct a burning mirror. You may try this – of course it is quite difficult to

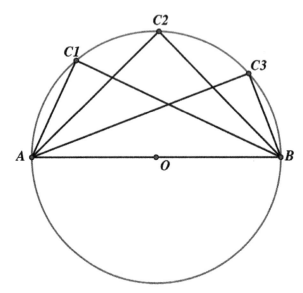

FIGURE 2.4   The theorem, known as one of Thales' Theorems, says that every triangle in a semicircle is a right-angled triangle.

make such a mirror – but at least you can see how a burning mirror would work if you look at the diagram from his treatise.

Before sharing this diagram, it would be useful to supply some background information. We will therefore remind ourselves of the mathematician who is credited with their invention, Archimedes. There is a legend of the account of the Roman fleet being set ablaze during the siege of Syracuse, a port in Sicily, between 215-212 BC, by the burning mirrors designed by Archimedes.[9] Archimedes was the native of this city and this was not the only time he was in trouble with the Romans, but the second time was the occasion when he was killed by a Roman soldier. He was apparently deep in his work on mathematics. As the advancing Roman army went through the neighborhood to evacuate the population, Archimedes asked not to be disturbed as he was too busy – only to be killed by the soldier in return. And a task for you, should you find yourself in Syracuse, would be to try to find the tomb of Archimedes. It seems like a simple task – we know he lived and died there and on his tomb there should or would have been an inscribed geometrical diagram, most probably one depicting one of his many contributions to geometry. Some say it is the diagram of a cylinder,

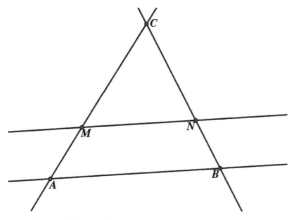

FIGURE 2.5   Another of Thales' Theorems: Is this the one which bears his name? It states that if a straight line is drawn parallel to one of the sides of a triangle, then it cuts the sides of the triangle proportionally; and, if the sides of the triangle are cut proportionally, then the line joining the points of section is parallel to the remaining side of the triangle. It appears also in Euclid, *Book VI, Proposition 2*.

used to calculate the volume of a sphere. This was an ingenious method and deserves an explanation. Say you have a sphere and want to calculate its volume. You know how to calculate the volume of a cylinder (area of the base times the height) and the volume of a cone (it is equal to the third of the product of base times height). What if you use these two formulae to find the third, the volume of the sphere? Archimedes came up with slicing up the concoction shown in Figure 2.6 and discovered that if these slices of cross sections moved, the cross sections of the cone and sphere at each point would exactly balance the cross section of the cylinder. And that is how he calculated the volume of the sphere: It is exactly 2/3 of a circumscribed cylinder.

Legend has it that Archimedes was so proud of this discovery that he asked for the diagram as given in Figure 2.6 to be inscribed on his grave. Imagine if he thought his Eureka moment (the first 'aha!' moment recorded in history) was a better discovery than this? I am sure you have heard of it – once he had a bath, Archimedes suddenly realized that the amount of water displaced as he sat in the bath was equal in volume to his body, giving a method for finding volume of more irregular shapes. The excitement of this thought made him walk out into the street naked, exclaiming 'eureka!'

To return to our burning mirrors, although there are the mentions of Archimedes' success in using them, it seems that Anthemius was the

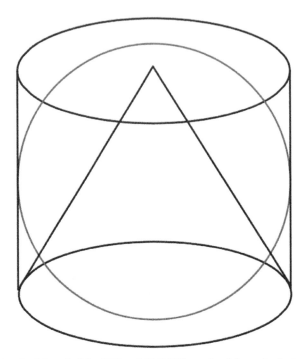

FIGURE 2.6   Archimedes' (c. 287–c. 212 BCE) method for calculating the volume of the sphere involved inscribing a sphere in a cylinder.

FIGURE 2.7   A photograph of a possible tomb of Archimedes' in Syracuse, Sicily, taken by the author.

first one to actually write how to make them, and explain the mathematics behind the idea. Anthemius was in his little book *On Burning Mirrors* quite specific and it seems likely that he tried at least several times to use them in practice: So how do they work? Anthemius said that one should use hexagonal mirrors hinged "with links and joints at the sides of larger hexagonal mirrors, each made to focus the sun's rays on the object",[10] and recommends that there should be no fewer than twenty-four such mirrors. The mirrors then, should focus the sun's rays on the object and the mathematical model for it would look along the lines of Figure 2.8.

I will continue to debate whether it was Anthemius or Archimedes who first designed burning mirrors, but if Archimedes did use them, as Anthemius mentioned, he didn't leave much of a description of how to make them. It is quite clear that Anthemius himself had to discover how to actually make these mirrors work. At least, he inherited a story that this can be done and that Archimedes was successful.[11] There expired, of course, quite a few centuries between them and another geometer named Diocles who also wrote a treatise *On Burning Mirrors* to which Anthemius might have had the access.[12]

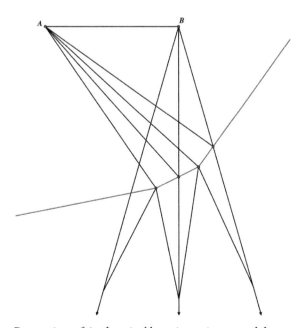

FIGURE 2.8   Recreation of Anthemius' burning mirror model.

ISIDORE, THE MILESIAN

The other architect of Hagia Sophia, Isidore, must have known of Thales too. After all, he was born in the same town from which Thales came. The Milesian school was a very important school that brought together knowledge that existed in Babylonia and Egypt, particularly that related to mathematics and astronomy, and made it part of the Hellenic intellectual heritage. Some famous philosophers and mathematicians also schooled there were Anaximander and Anaximenes. Their most important work was related to mathematics and to the rise of natural philosophy. The system they devised included a thought method like this: Make a case or a hypothesis about something, choose or devise a mechanism of operation, and then see whether you can prove whether your case or hypothesis works. The act of making hypotheses and then trying to see whether that worked became a successful tool that is still very much in operation. Although this was a beginning of the structure of mathematical and scientific model that will make Greek mathematics so important in the learning of mathematics for two millennia, Milesians weren't, at this time (around fifth to sixth century BCE) as important as Euclid and the Alexandrian school will become a couple of centuries later.

But nonetheless, Isidore came from there. It would be impossible that he would not be aware of such heritage. Isidore was also very much interested in the work of Archimedes and was the first one to compile what is being passed to us from Archimedes' work. Archimedes addressed his findings as letters which he sent to his contemporaries, some of whom worked at the Great Library of Alexandria. These survived long enough for Isidore to compile them around 530 AD and make a book out of them. In around 950 AD, a copy of Isidore's compilation was made by an anonymous scribe in Constantinople and traveled to Jerusalem where in 1229 this book, written on velum (leaves made of thin calf animal skin rather than papyrus) was scraped and a liturgical Christian text written over it.[13] We can therefore say with a level of certainty that Isidore used some of the knowledge he gained from Archimedes in his work on our great cathedral of Hagia Sophia.

APPROXIMATIONS IN MATHEMATICS, ARCHITECTURE, AND LIFE

Hagia Sophia's dome is an amazing structure – the first one built by Anthemius and Isidore did collapse only a couple of decades after it was finished (in 557), but the story of that is that the bricklayers used more mortar then brick in a rush to complete the structure and did not wait for the mortar to be fully set before applying the next level. This mistake took decades to become apparent. Additionally, the dome was probably too shallow. When the new dome was built by the nephew of Isidore, Isidore the Younger, he increased the arc and depth of the dome, and added ribs to provide further support for the structure.

This great cathedral, ultimately by Isidore and Anthemius, stood the test of time and many centuries of earthquakes. The dome was designed in such a way to create the greatest possible space uninterrupted by pillars or columns underneath it. It rests on four triangular pendentives which allow for the weight of the dome to be transferred to a square supporting the middle structure of the church. This, a very near square structure, is therefore covered by a circular structure – how did Isidore and Anthemius create this and what mathematics did they use?

Recent studies of Hagia Sophia show that they used many approximations for the construction of this great dome. For example, it is a well-known fact that Archimedes set π to be between 3 1/7 and 3 10/71, and this is what seems was used in the description of Hagia Sophia's circular shapes. Similarly, they seemed to have used the approximation of 17/12 to be equal to $\sqrt{2}$ to construct the central square's diagonal.

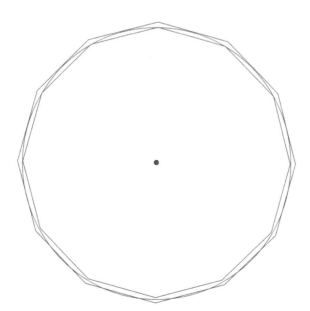

FIGURE 2.9   Archimedes' way of approximating π.

To find an approximation to π, Archimedes used a regular polygon with a circle inscribed in it, then another circle circumscribed around it all as in Figure 2.9. If he measured or calculated the perimeter of the polygon and kept increasing the number of sides, each time taking a new measurement, eventually the two circles would almost coincide, and from this he could work out the value of π as the ratio between a circle's circumference and its diameter.[14]

Such a magnificent structure as Hagia Sophia is then not completely accurate? It is as accurate as life gets and as close to the image of perfection as mathematicians that lived some fifteen hundred years ago were able to calculate. What is amazing is that they did not have calculating machines we have today, no internet to find things quickly or communicate with others, and had managed to do very precise work they had learnt from a mathematician who lived some seven centuries before them. Our Anthemius and Isidore, by careful preservation, adaptation, and when called for, approximation, created miraculous results which still stand today.

NOTES

1 Archimedes (c. 287–c. 212 BCE), Greek mathematician and scientist, believed by many to be one of the most influential thinkers of antiquity.
2 The burning mirrors of Archimedes have appeared in fiction, popular science, and academic publications as a common theme for some time. Some partial bibliography would be Clagett (2001) and Simms (1977).
3 Edited by L. Dupuy in 1777.
4 The history of Daras is rather unhappy. After many wars from the Middle Ages all the way to the twentieth century, it finally became a site of the massacre during the Armenian Genocide in 1915.
5 Isiodore of Miletus was most likely to be the author of a book which continued the tradition of Euclidean *Elements*. It is believed that he wrote a book which he called XV (the XIV having been written by Hypiscles, another late antiquity mathematician).
6 The five theorems attributed to Thales are:

- A circle is bisected by its diameter
- Angles at the base of any isosceles triangle are equal
- The angles between two intersecting straight lines are equal (later enlarged to be part of Euclid's Proposition VI.2
- Two triangles are congruent (equal in every respect) if they have two angles and one side equal
- An angle in a semicircle is a right angle.

7 The study was found by an eighteenth century French antiquary, L. Dupuy, who transcribed and published it in Paris in 1777 and can be easily found as electronic copy on the archive.org.

8 Euclid's *Elements* is a collection of thirteen books attributed to the Greek mathematician Euclid of Alexandria. (c. 300 BCE). This mathematical treatise is the oldest deductive treatment of mathematical sciences, providing a complete system of mathematics. Although that completeness has been questioned and successfully proven to have had a 'gap', giving rise to different types of geometries in the nineteenth century, it was still the most successful and probably the most influential mathematics textbook of all times.

9 Archimedes of Syracuse (c.287-c.212 BCE) was a Greek mathematician, engineer and inventor. I frequently mention him in this book and that should not be a huge surprise – he is considered by many to be the greatest mathematician of all time.

10 See Knoor (1983, p. 54).

11 A lot of the history of the geometry of burning mirrors has been described in Knorr (1983).

12 Diocles (c.240-c.180 BCE) was a Greek mathematician. Anthemius of Tralles (c.474-c.534) was born in Tralles, as his name testifies, today's Aydin in southwest Turkey. Isidore of Miletus (dates unknown) was born in Miletus, a Greek mathematician.

13 The story goes on but is a subject of an entirely different book, the *Archimedes Codex*, written by Reviel Netz and William Noel (2007).

14 Archimedes actually used the 96-sided polygons as in our diagram to find the approximation of π to be between 320/71 and 310/70.

CHAPTER 3

March

HIDDEN DOMES AND HEAVENLY IDEAS

There is another famous dome to that studied in February and quite different mathematics was invented as a consequence of its making. In March we discuss Christopher Wren, architect and mathematician, who died on 8th March 1723 and was buried in the Cathedral of St Paul in London on 16th March, the cathedral that he designed following the 1666 Great Fire of London. We look at the structure of the curve that was studied as a consequence of Wren's design, mentioning some contemporaries of Wren, such as Jacob Bernoulli and Robert Hooke. Newton remains fresh in our memory and we add the link to him and to the Bernoulli brothers, and Jacob's study of the Archimedean spiral. A famous mistake adorning the grave of this Bernoulli brother will take us to a doodle drawing in preparation for the month of April.

IF YOU EVER COME to London the visit to St Paul's is highly recommended, not only because it so clearly encompasses the story of this city, but more importantly of the people who have for centuries passed through it. It is an example of some beautiful mathematics and a monument to mark the beginning of the study of curves with extraordinary properties and strange names such as 'catenary'.

The cathedral is to be found in the very center of London, in the midst of the financial world. Here you can experience real hustle and bustle, so be prepared for it. When you see St Paul's remember this was the highest

FIGURE 3.1   Sir Christopher Wren (1632–1723). Line engraving by A. Bannerman after Sir. G. Kneller. Credit: Wellcome Collection. CC BY.

building in London from 1710 to the not so distant 1962. But the building itself signifies a few other things. It is perhaps the most significant landmark that marks the change of London from the medieval, feudal, and guild dominant city to a modern one that is at the center of global trade networks.

The founding of St Paul's Cathedral's began at the time of William the Conqueror in 1087 and by the seventeenth century it had became a center of spiritual as well as educational and civic life of London. Yet on the 2nd September 1666, a spark from a baker's oven in Pudding Lane, started a fire that raged in the city for four days. The intensity of the fire was such that a pottery melted – something that happens when temperatures reach above 1250C°. Whilst the fire was devastating in terms of the destruction of the inner fabric of the city, it is doubtful that many people died. It seems, in fact, that only six verified deaths were recorded among the seventy to eighty thousand Londoners who became homeless as a consequence of the event. The Great Fire of London is equally great in the psyche and the physicality of the history of London – it was a life-changing event for the city and probably for the whole region. Importantly for the history of mathematics, it was also the event that brought about the rebuilding of St Paul's Cathedral.

The entire process of rebuilding lasted from 1675 to 1720. The fire, which destroyed so much of the city, changed also the building trade and how it operated in the city until this moment. The masons' guilds of London had an elaborate system of trade control over the regions in which they operated but the need for rebuilding the city fast and efficiently meant that the existing controls had to change and relax in order to attract and maintain a sufficient number of builders. At the same time the division of roles in the trade had to change – there was suddenly a lack of sufficiently skilled builders and so a new group emerged, that of professional architects. An architect has always been a recognizable role but now the number of architects became greater and their roles diversified. From one general main role of architect as a designer, there sprung at least three more: the architect supervisor to oversee the building process, the architect who knew how to do things, such as a master mason, and the architectural writer and critic. All of these became viable roles within the architectural trade.[1]

Sir Christopher Wren came to this scene from Oxford when his own practice was just only taking hold. Whilst he remained close to builders with whom he worked, he was very much a new type of the architect too. His ideas were, above all, based on mathematical principles, and a reflection of his interest in connecting the two – architecture and mathematics.

Wren was one of the founding members of the Royal Society and in 1661 became the Savillian Professor of Astronomy at Oxford.[2] In the same year he was appointed to work on the overdue repairs of the old St Paul's.[3] Wren finished plans for repairing the dome of the old cathedral and his proposals were accepted on 27th August 1666. As we now know, only a week later the Great Fire reduced the whole building to rubble. The design process could therefore begin afresh and the whole new structure could be built. As a consequence, the redesign could now have a completely new approach to the cathedral and the dome.

The dome of St Paul's is a multilayered experience, as the dome you see outside is not exactly the same dome that you see inside the cathedral. The dome in fact is a three-dome nested structure. The inner dome is the lowest of the three and its cross-section is a simple hemisphere. The middle dome is nested between the upper and lower domes and gives the strength and structure for a lantern to be placed in the external dome. It is a widely held popular belief that Wren used his knowledge of a catenary curve to make this dome, but we know that the truth (as usual) is not as simple as it first appears.

FIGURE 3.2   Wren's sketch of the design of the dome showing how different curves were used to model the nested domes of St Paul's Cathedral. © The Trustees of the British Museum.

THE HANGING CHAIN CURVE

Together with Robert Hooke[4], a friend and collaborator, Wren worked on a curve that would withstand the greatest possible tension. Wren and Hooke realized that a curve such as catenary would be perfect for building domes or arches. A catenary is a curve that models the hanging chain. It has a shape that an ideal chain would assume under its own weight and hanging only from its two ends. The name comes from *catena*, which is Latin for chains. It is a curve that is very similar, at its summit, to a section of parabola, a cubic parabola, and even a circle.

Hooke wrote a paper about this curve but knowledge always being equal to power of one kind or another, he did not want to divulge the secret of his finding. He was still only at the beginning of the discovery of this curve and did not have the equation for it. But he was sure that a curve that models a perfect arch is a curve of a hanging chain, so he hid the information in an anagram, which appeared in one of his papers. The solution to this secret remained safe until Hooke died and the executor of his will provided the full text in Latin in 1705: *Ut pendet continuum flexile, sic stabit contiguum rigidum invesrum* (As hangs a flexible cable so inverted stand the touching

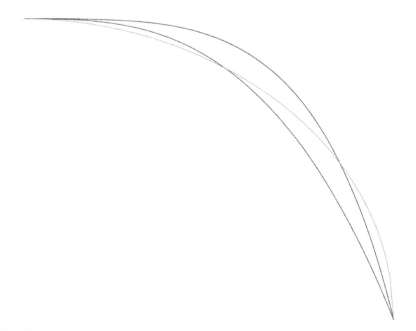

FIGURE 3.3   Catenary curve (red) compared with the sections of a cubic parabola (blue) and a circle (yellow).

pieces of an arch). The beauty of this simple statement is quite staggering. This knowledge has enabled many architects and engineers since to successfully use the principle in various architectural projects.[5] But the first and most prominent would be the dome of St Paul's, although Wren only managed to get an approximation to the actual shape of catenary curve. Wren attempted to come up with its mathematical description but was unable to do so. A useful and never underused architectural and engineering skill came handy here too: He realized that a very close approximation may be almost as good as the real thing. He therefore used a section of a curve known for centuries, the simplest form of cubic, $y=x^3$. This curve, when a section of it is rotated, makes a cubic paraboloid, which is the shape of both the top and middle dome of St Paul's Cathedral.

Before the Internet, search engines, and social media, scientific news still managed to travel – perhaps not as fast as today, but it traveled nonetheless. Mathematicians from around Europe corresponded and exchanged ideas and often posed to each other their mathematical questions, problems, and challenges. In this way, a mathematical description came about for the catenary curve. First of course the question had to be posed and the problem formulated. This was done by Jacob Bernoulli when in 1691 he invited people to submit their answers so that they could be published in the June issue of the journal *Acta Eruditorum*, the leading scientific journal of the German lands in Europe at the time, published in Leipzig between 1682 and 1782. Two mathematicians provided their answers: Johann Bernoulli (brother of Jacob) and Christian Huygens.[6]

Without going into detail of these scientific papers, the mathematical description of the curve was an important advance in mathematics, showing the analytical nature of the curve, providing knowledge about its properties, and how it can be geometrically constructed. Whilst the papers were important in their own right, they also touched upon an issue of particular interest to mathematicians of the time. They were keen to discover more about the nature and precise mathematical descriptions of motions that can be studied through curves.[7] Such curves would be able to shed light on the orbits of planets. They would demonstrate the structure of an order that underlies every motion and provide an understanding of the pathways of planets. Perhaps these mathematicians were interested in this above all: What kind of universal order is there? And how can this be known, understood, and studied mathematically?

The study of the catenary curve led to a few more interesting findings. One of these is the description of brachistochrone curve. A brachistochrone

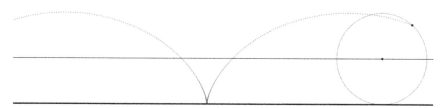

FIGURE 3.4   How a cycloid is generated. The brachistochrone is the section of a cycloid.

is a curve that provides a line of fastest descent. This means that an object rolling down an inclined brachistochrone curve will reach the lowest point faster on this curve than if it rolled on an inclined straight line.

To make your own brachistochrone curve, you would have to know the principles upon which this curve is constructed. Imagine a circle that rolls in a uniform fashion on a straight line. One of the points on its circumference, as the circle rolls ahead, traces a curve. This curve is called brachistochrone – 'brahistos' meaning the shortest and 'chronos' meaning time in ancient Greek.

This curve was described by Fermat and by Galileo before, but it was the Bernoulli brothers that set to work on this curve and provide its algebraic description.[8]

WREN AND OTHER CURVED SHAPES

Wren made some mathematical discoveries that were not to be applied in London or any other cityscape for centuries to come. One is a geometrical object bound by a curved surface, called cylindrical hyperboloid, which has often been employed in modern architecture.[9] The curved surface of cylindrical hyperboloid is made by the movement of a straight line segment whose end points are to be found on two circles, with both lines moving at a uniform speed. There is a whole family of hyperboloid surfaces but Wren described this, a one sheet hyperboloid as in Figure 3.5, and showed it to be a ruled surface.

A surface is said to be ruled if, through every point that lies on it, a straight line can be drawn that also lies on the surface without pearcing or protruding. This does not mean that we can draw *any* straight line as we like it (try imagining how you can do this on the cylindrical hyperboloid), but it is important to understand that there is *at least one such line* that can be drawn through each point of such a surface. Wren's paper was presented to the Royal Society in 1669 and the anecdotal tale suggests that Wren came up with the idea about describing and defining ruled surfaces

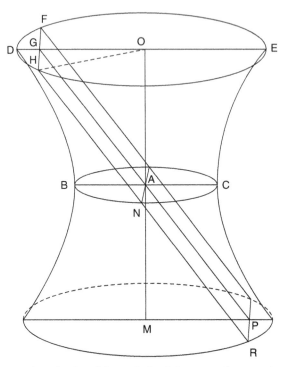

FIGURE 3.5 Wren's cylindrical hyperboloid diagram showing how it appeared for the first time in the paper "Generatio Corporis Cylindroidis Hyperbolici, etc." in *Philosophical Transactions of the Royal Society,* 1st January 1669, vol. 4, London.

by seeing a round wicker basket in a window shop walking around Oxford. He noticed that

> *If one of the osiers led around the axis of a cylinder and preserving that oblique position with respect to the axis would describe that concavo-convex surface, and so cylindroids of that sort could be made on a lathe by means of a straight steel tool in a position oblique to the axis of the cylinder, the section of which through the axis will be that curved line.*[10]

BEAUTY IS IN THE EYE OF THE BEHOLDER – OR IN MATHEMATICS?

Wren worked in architecture, discovered new mathematics, and was an accomplished astronomer. To a modern person this range of activity and knowledge may appear at first very broad. These disciplines require good application of skills and knowledge, not to mention the focus to complete the necessary research, but to Wren these were disciplines that combine the pursuit of finding the same principles and reveal to us the nature of the world and the order through which it was made. Perhaps this is best seen through his writing about beauty:

> *There are two Causes of Beauty, natural and customary. Natural is from Geometry, consisting in Uniformity (that is Equality) and Proportion. Customary Beauty is begotten by the Use of our Senses to those Objects which are usually pleasing to us for other Causes, as Familiarity or particular Inclination breeds a Love to Things not in themselves lovely. Here lies the great Occasion of Errors: here is tried the Architect's Judgment: but always the true Test is natural or geometric Beauty.*[11]

And hence the explanation of his interest in connecting the work in mathematics (mainly geometry), astronomy, and architecture. They all led him to make small steps towards understanding those principles of beauty that lie in the geometrical principles that can be found, tested, and understood in the natural world, and recreated in our built environment.

SOME OTHER INTERESTING CURVES

The Great Fire of London was officially commemorated by two monuments to mark its beginning and end. The first, the monumental column made in the classical Doric style, is simply called *The Monument* and is positioned at the point where the fire started – just west of the Pudding Lane, at the junction of Fish Street Hill and Monument Street. The second is a smaller monument that celebrates where the fire stopped with a wooden figure of a boy covered with gold, standing in the corner of a building at the corner of Giltspur Street and Cock Lane in Smithfield, central London.

The Monument is more interesting of course, not only because it is larger but because it too was designed by Christopher Wren and his friend, scientist, and Royal Society member, Robert Hooke. The column is a hollow structure with a spiral staircase that leads to the top terrace, which can still be visited today and where you can observe the London cityscape from its viewing platform. This spiral staircase is also an interesting thing – it celebrates not only the new city that London became after the Great Fire but the study of curves to which Wren contributed, and which in turn lead to some other curves and inventions.

We should not forget that the study of curves increased, as we saw, around the time Wren was active. The spiral was also one of the curves that was revisited. A curve of an eternal appeal, spirals were studied since antiquity. There is of course not only one possible spiral. Try with some friends and see how many different types you will come up with. In terms of mathematics of spirals though, and the mathematics of Wren's time, we can simplify matters and look at two types of spiral and how they are constructed.

At this point we meet with Archimedes again. The first mathematical description of a spiral is known as Archimedean spiral in his work *On Spirals*, towards the end of his life in 225 BCE. It surprised me when I first came across this that he used quite a modern description of how to create this curve – *kinematically*, that is, through a movement. The curve, he said, is an intersecting moving point of the two constraining principles:

> *If a straight line one extremity of which remains fixed is made to revolve at a uniform rate in a plane until it returns to the position from which it started, and if, at the same time as the straight line is revolving, a point moves at a uniform rate along the straight line, starting from the fixed extremity, the point will describe a spiral in the plane.*[12]

Archimedes used this delightful spiral to provide a demonstration of a trisection of an angle and squaring of a circle, two famous problems of geometry from antiquity. This second problem was proved to be an impossibility in 1882.[13]

Fast forward many centuries and playing on this spiral, the French philosopher René Descartes in 1638 described the logarithmic spiral. This curve is given as our second spiral, shown in the diagram in Figure 3.7. The difference between the two spirals – Archimedean (Figure 3.6) and logarithmic (Figure 3.7) is that the successive turns of logarithmic spiral increase in geometric progression but not in the Archimedean spiral. The logarithmic spiral is also called equiangular, meaning of equal angles, as every ray from its center intersects every turn of the spiral at an equal angle.

Jacob Bernoulli, who we already met a little earlier, was inspired by all this talk of curves and inspired others to think of them. You will remember he invited mathematicians to write for *Acta Auditorum* and give the answer to how a curve which models a hanging chain can be formulated and described mathematically. Jacob himself became so enchanted by the logarithmic spiral, that he called it *spira mirabilis*, the miraculous curve, and

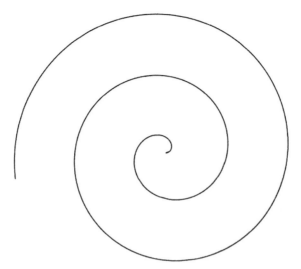

FIGURE 3.6   The Archimedean spiral.

FIGURE 3.7   The equiangular or logarithmic spiral.

asked for it to be engraved on his tombstone, with the phrase 'Edem mutate resurgo' – I shall arise the same though changed.

But what happened? The stone-mason didn't know the difference between the spira mirabilis and Archimedean spiral? He didn't have anyone to tell him that he made a mistake? We can't tell, but the spiral that appears on his gravestone is not his spira mirabilis but Archimedean spiral – see Figure 3.8. To me it seems unlikely that Jacob would leave such an important thing as making the spiral for his eternal resting place to be engraved without his supervision. On the other hand, he may have already passed and was unable to supervise its execution. Some think that this is one of the most famous mistakes in the history of mathematics.[14]

FIGURE 3.8   Jacob Bernoulli's tombstone in Basler Münster, Switzerland.

Whatever the reason, the spiral that marks Jacob's grave is Archimedean and not the one he so cherished, through which he would arise again though change. It may seem that we go on eternally in spirals but I must mention briefly again that Wren instead managed to have his dream of resurgence survive to this day. St Paul's fashions a carved phoenix at the portioco of its south door, marked with a similar phrase 'resurgam' – 'I shall rise again'.

NOTES

1 John Evelyn (1620–1706) divided the profession into four main roles in his *Parallel of the Ancient Architecture with Modern*, published in London in 1664. He was a founding member of the Royal Society and an author on various subjects related to architecture, fine arts, forestry, and religious topics, publishing about thirty books within his lifetime. Evelyn's description of the four types of architects in 1664 are worth pursuing in our investigation of the origin of the concept. They were:

- Architectus Ingenio – the one who designs the building
- Architectus Sumptuarium – the one who superintendents and presides over the works; the client who pays for the building: this may be a patron but also his representative
- Architectus Manuarius – master mason, etc.
- Architectus Verborum – architectural writer.

2 Christopher Wren (1632–1723) is often referred to as the greatest English architect of his time.

3 In the next five years, Wren designed the chapel of Pembroke College in Cambridge and the Sheldonian Theatre next to the Bodleian Library in Oxford.

4 Robert Hooke (1635–1703) was an English philosopher, architect, and scientist.

5 How knowledge spreads across boundaries is always an interesting phenomenon – I very much recommend reading how Hooke's Chain Theory influenced the construction of catenary arches in Spain during the eighteenth century. See Ginovart, et al. (2017).

6 Jacob Bernoulli (1655–1705) was also known as James or Jacques. One of the prominent mathematicians in the Bernoulli family, he was a proponent of Leibnizian calculus. He was brother of course of Johann Bernoulli (1667–1748) who was in turn also known for his contribution to infinitesimal calculus. Christiaan Huygens (1629–1695) was a Dutch physicist, mathematician and inventor.

7 An example would be Gottfried Wilhelm Leibniz (1646–1716), a German mathematician and philosopher, who was interested in continuing the work of Kepler and his contemporaries, such as Bernoullis and Huygens.

8 Pierre de Fermat (1607–1665) was a prolific French mathematician, but officially a lawyer. He also worked on the curve, as did Galileo, who mentioned brachistochrone in his *The Two Sciences* (1638), writing in his Third Day Theorem 22, Proposition 36: *From the preceding it is possible to infer that the quickest path of all from one point to another, is not the shortest path, namely a straight line, but the arc of a circle.*

9 Two examples of the use of cylindrical hyperboloid in modern architecture are for example Saint Louis Science Center's James S. McDonnel Planetarium (built 1963) and Newcastle International air traffic control tower, Newcastle upon Tyne (built 1967).

10 Mentioned in Lehmann (1945), p. 1.

11 Wren, *Parentalia*, Tract I (1750: 351).

12 This quote is given in Heath, (1921), p. 64.

13 The three problems of antiquity, the squaring of the circle, trisection of an angle, and the doubling of cube are wonderfully described in Hobson's *Squaring the Circle: A History of the Problem* (1913), Cambridge University Press.

14 See full story about this in Yates (1952).

CHAPTER 4

April

AN IDEAL AND EMMY

To doodle or not to doodle – and is there an ideal doodle or an ideal spiral for that matter? Those are the thoughts that interest some mathematicians. The idea that mathematics can lead to or that it may be the source of poetic beauty is not only the case when it can be found, recognized or applied in something made into a concrete object, as through architecture. Such beauty can be found in a very abstract way of doing mathematics and through very abstract concepts. In April we could do much worse than to think of Emmy Noether (1882–1935), a German Jewish mathematician and inventor of modern algebra, who died in April of 1935. Noether is often called the 'mother of abstract algebra'. She developed the general theory of ideals and Noetherian rings.[1] How something so abstract can be so beautiful is the main question of this chapter.

AMALIE EMMY NOETHER MASTERED three languages as a girl and was her father's mathematical conversation companion, probably for most of the time they shared together. He was a mathematician himself, working at Erlangen, Bavaria from 1875, only three years after Felix Klein[2] introduced his program for the study of mathematics. This was an approach that made a huge difference to the study of geometry. In Euclidean geometry, for example, lengths and angles are unchanged under rotations. But what happens in other geometries?

FIGURE 4.1 Emmy Noether (1882–1935). Source archives of P. Roquette, Heidelberg and Clark Kimberling, Evansville. Reproduced by the kind permission of Niedersächsische Staats- und Universitätsbibliothek Göttingen.

Klein's famous program pointed to the 'possibility of new geometries defined from diverse transformation groups'.[3] It began new investigations in geometry based on projective geometry and group theory. Projective geometry studies geometric properties that are unchanged through various projective transformations, whilst, as the name implies, group theory studies groups. Groups are mathematical sets within which mathematical operations that satisfy certain basic properties are valid. Groups have a fundamental connection with geometry – and with the notion of symmetry. In geometry for example, an object is assigned a symmetry group which consists of the set of transformations under which that object is unchanged, or invariant.

NOETHER'S ABSTRACT ART OF MATHEMATICS

I have a task now to explain what Noether is famous for other than the fact that she was a woman mathematician in the early twentieth century. To start from the very basic description of herself as a 'mother' of abstract algebra, how do we explain what abstract or modern algebra is? We may well use an analogy here of the abstract painting. In a classical painting, there is a strict adherence to the rules that govern presentation of space, colour, and styles that are used by the artist, each with an aim to create a scene that the painting presents to us. Even if the depiction of whatever the scene tells us is sometimes puzzling, it is more or less true to what we would expect to see in real-life. With abstract painting the rules are different. But can we say, a century or more after this too has become a norm, that such works do not follow their own rules and do not depict something from real-life? This is debatable but the question remains: What is *reality*? The *real thing* may not only be a person but a feeling or a thought, and abstract art often tries to depict these. Here however, lies my difficulty in explaining the beauty of Noether's mathematics, as many people already find ordinary school-based algebra abstract!

Perhaps an art theorist could help to illustrate the comparision further. Guillaume Apollinaire,[4] the man who coined the term cubism in 1911, orphism in 1912, and surrealism in 1917, gave us a great rule to create modern art that changed the history of art.[5] Modern art, in contrast to classical art, has a crucial characteristic:

> *Real resemblance no longer has any importance, since everything is sacrificed by the artist to truth, to the necessities of a higher nature whose existence he assumes, but does not lay bare.*[6]

If we were to translate this to mathematics, a question would certainly be: How did Noether take algebra from classical to modern? The task of making something as abstract as algebra and making it more abstract 'in a new way, albeit a deeper, more significant way'[7] is how her work has been described. Let us explain this with yet another illustration and approach.

Noether's most famous work is probably her *Idealtheorie in Ringbereichen* (*Ideal Theory in Rings*).[8] This is her work on 'ideal theory' – it showed how an 'ideal' can be considered as the least common multiple of irreducible ideals or the least common multiple of maximal primary ideals, and so on. Perhaps this is something on which we can work. If we compare it with painting, again, the basis of a painting can certainly be considered to be colour. Similarly, in mathematics, to make a simple comparison, the essence of operations or properties of mathematical elements can be studied themselves. Let us 'zoom' in ever closer and make another comparison. Everyone should know what an *ideal* is, as much as anyone with sight could say what *blue* is. To use our art comparisons, instead of painting an abstract world using only colours, Noether in her work and in *Idealtheorie* managed to paint an abstract world of mathematics that is explainable by mathematical *ideals*. An *ideal* in abstract algebra is a special subset of some ring (a fundamental structure in abstract algebra, consisting of a set that has two binary operations that generalize addition and multiplication), something that fundamentally defines that ring. We'll come back to it again shortly.

Consider modern art and architecture of the twentieth century and how such objects appeared to the collective eye of societies that were, until then, accustomed to classical ideals. There was an admiration, a renewed interest and sometimes a backlash of this new approach in cultural contexts. None of those reactions have yet disappeared from the discourse of differences between modern and abstract in general. The modern seemed to break all romantic ideals – it itself became an ideal. In mathematics, similar to art and architecture, the emphasis on new forms became overwhelming in the same period. The structure for creating new mathematics was nevertheless rigorous, although the criteria for recognizing and appreciating the values and the beauty of these creations perhaps became less well known. We can now make an inverse comparison with modern art or architecture: There too the process of creation through an underlying system of values was also rigorous, but in its own way.

Perhaps the best illustration of how this is done is by looking at the work of one of the fathers of modernism in architecture, Le Corbusier.[9] Corbusier built on the tradition of Vitruvius and gave a modern interpretation of the *Vitruvian Man*, which sought to universally describe and standardize measurements of the human body so that they could be used in standardizing architectural design. The *Vitruvian Man* sought to show a creature whose proportions and image we more or less can all universally admire, and which is so beautifully illustrated by Leonardo's famous image.[10] Le Corbusier's man, on the other hand, was not a beautiful creature like Leonardo's at first sight, but it nonetheless was very useful in giving architects the most important ratios in the measurements of an average human form.[11] These could then be used as a system of measurements to constrain and inform the design of a building. For classical architects, the beauty of the human form was linked to the beauty of the architectural edifice, which was to contain it when built – and the symmetries of both would inform the process of design. However, for Corbusier, the beauty was to be found in the functional complexity of the built environment, producing through architectural design a *machine for living in*. Whether we experience the beauty of modern architecture (especially that which relates in some way to our example of Corbusier's system) depends in some way on whether we see beyond the structure in front of us, and whether we can understand its underlying principles. A process of imagining the underlying structure of a building and how that influences the senses and produces the aesthetically pleasing effect objects may have on us, is very much left to the observer. This is perhaps the main feature of modern architecture and art, in comparison with the classical, where this process of producing an aesthetically pleasing experience is much more under the control of the artist or architect themselves.

The creative act in addition to the aesthetic experience are both part of what can be described as a double act:

> *In the moment of appreciation we live again the moment when the creator saw and held the hidden likeness... When a theory is at once fresh and convincing, we do not merely nod over someone else's work. We re-enact the creative act, and we ourselves make the discovery again.*[12]

NOETHER SET DOWN THE LAW

Now we can look again at *ideals* in mathematics (in abstract algebra). Similar to what we witnessed in architecture, there are main characteristics of a human that we can idealize: they are not of course same for every human and probably not even close to some. But in these 'ideals' are contained all crucial information and characteristics of a human body and its movements, so that architects can design a space that most people will be able to use.

In mathematics, too, we look for some *ideals*, and in abstract algebra it has a precise meaning: Ideals are used to portray and generalize certain subsets of objects with their most important characteristic. The set of even numbers, for example, or multiples of a number, are examples of such ideals. Their property is that they preserve this most important characteristic (remain being even) through certain processes or operations. Even numbers also preserve the property of being even after they are multiplied. The concept of the theory of ideals, if you wish to actually do it, is much more complicated than that, but so is making beautiful architecture!

Noether also came up with a theorem which now bares her name: Noether's Theorem. We need to briefly recapture the concept of symmetry first before exploring this theorem. If you can, fold a picture of something and if the two halves coincide, this means that that object has a reflecive symmetry. There are other symetries too: moving an object or a system by translation or rotation are also symmetry transformations. The mathematical meaning of symmetry is that after putting an object through such a transformation, the object (or system) remains the same, unchanged,

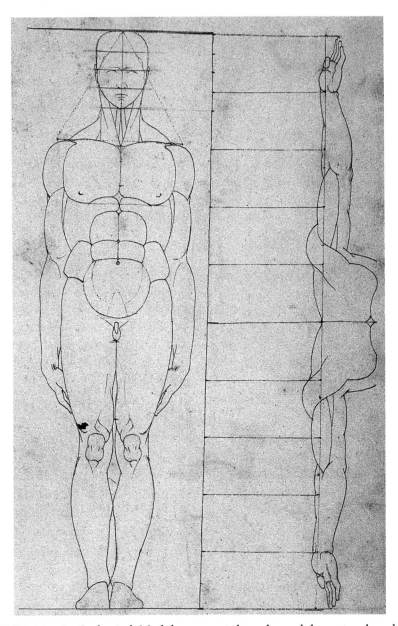

FIGURE 4.2   Le Corbusier's Modulor was not the only modular system based on human proportions – there were many in between *Vitruvian Man* and the Modulor; here is an anonymous artist's depiction from the nineteenth century. Human proportions: two figures. Drawing, 18–. Credit: Wellcome Collection. CC BY.

invariant. Noether studied what happens with systems that are put under such transformations in order to understand their behavior. Her theorem states that systems that remain unchanged through certain types of symmetries satisfy certain conservation laws of nature. The theory is highly abstract (as are other things Noether worked on), but perhaps it can be summarized as a set of tools to determine what quantities are unchanged from the symmetries of a system. For the avid student of higher mathematics there are resources that can be recommended from which one can further their study of this Theorem,[13] but here we will be satisfied to say that this, a most beautiful theorem, is so important because of its universality, the connection between such important concepts of symmetry and conservation, and its wide applications in other branches of mathematics, such as mechanics, field theory, and digital science, to name but a few.

WHAT ABOUT THE IDEAL AND EMMY?

Noether faced many aspects of discrimination during her life. First as a woman and then with the rise of the Nazi movement, as a Jewish woman. Nevertheless, by all accounts, she faced these difficulties with steadfastness and determination. In 1915, Noether was invited by fellow mathematicians David Hilbert and Felix Klein to join the mathematics department at the University of Göttingen, yet although there were objections that women can't possibly teach there, she spent four years lecturing under Hilbert's name. Then, in 1933 the lecture halls became closed to those of Jewish origin. She continued teaching at her own home and it is reported that even though some of the young students wore brown shirts – a sign of the Nazi party's paramilitary wing – they were welcome as students and she taught them without fear and with the hope that the human spirit cannot withstand evil for too long. It is quite unbelievable to think of this now, but Noether even went to Göttingen to work with her university colleagues for the summer vacations in 1934, a year after she had found asylum in the United States.[14]

Noether died in less than two years after her exile to the United States. There she had a position at Bryn Mawr College and had enjoyed settling in to this new life. But a minor health issue that took her to hospital for a routine and small operation resulted in her early death. One of the forebearers of modern mathematics, Noether managed to work her way up to the very top of the mathematical profession in a very difficult era, considering the context of her own heritage and the times in which she lived.

Let us then look around at how much has changed since those times. Women are of course given full rights in the developed world, but even so women mathematicians are a rarity. In 2013, women made only 6% of all professors of mathematics in the UK.

Noether's work is a fantastic example of how modern mathematics differs from classical, with good analogies to be found in both art and architecture from that time. The colours and shapes of modern art and architecture seemed to relentlessly break the romantic ideas of the past and their emphasis on new forms was overwhelming. Noether's mathematics similarly tells the story of modern mathematics in the making. And Emmy herself offers an *ideal* image of a woman mathematician fighting against the odds that her sex and origin gave her, to become the mother of modern algebra. Her friend and colleague gave an elegy at her funeral and described her as a 'primordial human rock', rather than 'clay, harmoniously shaped by God's artistic hand', a fitting tribute to an abstract thinker whose genius transcended all the limitations of her immediate reality.

NOTES

1 On the meaning of these terms we will talk about more later in the chapter, but if you want the full story, see Tent (2008); Moore (1967); Neuenschwander (2010).

2 Felix Klein (1849–1925) was a German mathematician best known for his Erlangen Program and for his work in non-euclidean geometry. The Erlangen Program is based on connecting classical geometry with group theory and projective geometry. A recommended reading is by Ji et al. (2015).

3 See Neuenschwander (2010, p.18).

4 Guillaume Apollinaire (1880–1918), was a French poet, playwright, and art critic of Polish-Belarusian descent.

5 Cubism and surrealism are now well-known art movements, but orphism was also an important movement in its own right at the beginning of the twentieth century. Unlike the abstraction of space and time or context, it focused on abstraction of pure colour. An example of this can be seen in the work of Robert Delaunay.

6 Chipp, H.B. (1968), p. 222.

7 Gray, J. (2008), p. 665.

8 Noether, E. (1921).

9 Charlers-Édouard Jeanneret, also known as Le Corbusier (1887–1965), was a Swiss-French architect, designer, and urbanist.

10 Leonardo da Vinci's *L'Uomo Vitruviano*, drawn around 1490.

11 Some recent authors have pointed out the quite misogynistic way of interpreting the human body through a male figure, rather than a female one, and designing the built environment around such canon of proportions. In defence of Corbusier, I must state that his Modulor, both in name, and in shape, can safely represent pretty much a body of any gender or race, unlike the Vitruvian man of Leonardo.

12 Bronowski (1956), p. 19.

13 See Neuenschwander (2010).

14 Hermann Weyl's speech at Noether's funeral, available in full at www.rzuser. uni-heidelberg.de/~ci3/weyl+noether.pdf.

CHAPTER 5

May

THE WITCH THAT TURNED

Mathematics can be misunderstood! It can also be translated in a wrong way. In May we turn to Maria Gaetana Agnesi, the first female professor of mathematics, who was born on 16th May 1718. In her famous book on mathematics, Agnesi summarized the principles of calculus and introduced to the wider world a curve that was called la Versiera (the turn). In English, this curve became known as the Witch of Agnesi through an erroneous translation. The curve, some of its amazing (later discovered) applications, and its appearance, will give us an impetus to travel on a pathway around some imaginary and real landscapes, investigating the transformative role mathematics had on society of the times.

IN THE PREVIOUS CHAPTER, we learnt a little about Emmy Noether, whose work on abstract algebra was ground-breaking for the development of modern mathematics. But she was not, of course, the first female mathematician to teach at a university, despite being the first woman to teach mathematics at the University of Göttingen. In fact, the first professor of mathematics to hold a university position was Maria Gaetana Agnesi.

As a young child Agnesi enchanted the literati of her beautiful city of Milan, by her talks and interpretations of contemporary mathematics. During the gatherings her father organized at their home, she was able, as a young girl, to make scholarly conversations in Latin. Agnesi became the first professor of mathematics, having been appointed to the Chair at the

FIGURE 5.1   Maria Gaetana Agnesi (1718–1799). Line engraving by E. Conquy after M. Longhi. Credit: Wellcome Collection. CC BY.

University of Bologna in 1750, although she never gave a lecture there. She was the first woman to publish a book on mathematics, an early systematization of calculus that remained in use for many years.

NEWTON FOR, AND BY, LADIES

A couple of years ago I was invited by an international women's charity to give a talk on the history of mathematics. The venue and the organization's branch, was in Canterbury, England. The cathedral, as I discovered only a couple of months prior to this invitation, has an extensive archive and a library with some extraordinary collections. One such collection was a bequest from an eighteenth century physician to his local parish church, of books mainly related to science. This wonderful collection, although not huge in the number, contained books that were incredibly valuable. Until the 1980s the books from this collection were to be found in his parish church's entrance hall, but luckily for the preservation of intellectual history, they were brought to the cathedral and put on the shelves of its library. Among this collection you can find such publications as the first edition of Euclid's *Elements* in English language, and the first edition of Newton's *Principia*.[1] It was on this book I wanted to concentrate for my talk, but as the organization was female, I also wanted to pay homage to a mathematician who developed Newton's ideas further and made them accessible, and that was how I decided to talk about Agnesi.

I settled on the title *Newton for Ladies*. Why suggest that Newton for ladies is any different from Newton for gentleman or otherwise everymen or everywomen? This is precisely why I decided on my title, as, whilst in our time *every* and *all* should be treated in the same way, in the eighteenth century it was not so – ladies, gentleman, and Newton, were treated quite distinctly. This in itself, apart from the mathematics I wanted to discuss, was a lesson worth remembering in order that we can celebrate what has been achieved since.

THE BOOK

The title of my talk was, known by those of you who haunt the British or other libraries for old books, a title of the little book written in the eighteenth century. *Newtonianism for Ladies* was published in 1737 in London and written by Francesco Algarotti,[2] one of the first 'beaux esprits' of his age. He was born in Venice and lived in many cities across Europe: He moved to Rome, then Bologna, and lived in Paris, London, St Petersburg, Krakow,

Dresden, Berlin, and Pisa, with his final destination being Venice where he died. Algarotti made friends with scientists in these places. In London he became a Fellow of the Royal Society, whilst in France he became a friend with Voltaire and his mistress Madame du Châtelet, and in Prussia he made friends with Frederick the Great.[3] Frederick the Great even bestowed a title upon him, and so Algarotti became a Prussian Count. In Krakow, Algarotti made friends with the Emperor of Poland, Augustus III.[4] How did he manage all this? Algarotti seemed, by all these accounts, to had been a beautiful, inspiring and an inspired spirit who felt at home everywhere he visited. But to have grown into such a free and charming spirit, he most probably owed to the study he undertook when he was young.

When he was only twenty years of age, Algarotti wrote *Neutoniasmo per la dame*, originally published in Italian then translated into English and published in England in 1737 as *Newtonianism for Ladies, or Dialogues on Light and Colours*. The book became a bestseller and was republished many times in the eighteenth century, and translated into many different languages. It adopted a model of teaching that had a form of a dialogue between a teacher and a learner, as was often the case at the time.[5]

THE SPECTACLE OF NEWTONIANISM

The book is divided into six dialogues taking place over two days, between a chevalier, the teacher, and a marchioness. As marchioness and the chevalier walk around an imaginary villa near Lake Garda in the most beautiful surroundings, they walk through various rooms of the building and scenes that offer not only a backdrop to the scientific discussion but a theatre of memory, which a reader of the book may be able to recreate for themselves by remembering the scenes.[6]

In the first dialogue the chevalier reads a poem that celebrates Bolognese professor Laura Bassi, a confirmed Newtonian philosopher who graduated from the University of Bologna just a few years earlier in 1732.[7] The poem contains a reference to the complex nature of light (remember Newton and his light prism? You have probably seen that in many images of Newton), so the marchioness stops the chevalier and asks about it. There it is – an introduction to Newtonian philosophy and science and of course, without much effort, an introduction to the world in which women can learn, be enlightened, and become part of the enlightenment. Of course, all of this is happening because of a helping hand of a chevalier who undertakes to initiate the unknowing. Here is, for me, the primal scene of what we nowadays call 'mansplaining' of a dilettante teaching women about science

FIGURE 5.2   Algarotti's grave in Pisa. Photo credit: author.

and philosophy and being put on the same pedestal as the groundbreaking scientists such as Bassi and Agnesi. In the eighteenth century, however, this was a progressive step and therefore it is quite forgivable.

The *Newtonianism for Ladies* was probably one of the main transporters of Newtonian ideas around Europe as it offered light-hearted but scientifically sound explanations of some of the principles of Newtonian science. It influenced many intellectuals across Europe in its time. Scientific theories are not usually so well known or popular, but we can safely say that Newtonian science, not only his Law of Universal Gravitation, or his contribution to the development of mathematics, was as popular (or more) as the General Theory of Relativity was in the twentieth century. Newtonian science offered a different view of the world, a view in which scientific exploration led us to discover laws that underlined the structure of the world. We must not forget though, that Newtonian science was set against other systems of sciences that also tried to unify the theory of scientific knowledge similar to, for example, his arch-enemy Leibniz's science.

But Newtonian science has unified many aspects of the discovery, manipulation, and communication of knowledge through scientific experimentation and mathematical exploration that had a unique attraction for certain intellectual circles in Europe. And in some strange way, it attracted many literary and philosophical friends such as Pope, Desaguliers, and even Voltaire, as we saw in Chapter 1.[8]

AGNESI AND THE WITCH

Agnesi was one such admirer but also a contributor to Newtonian science. As a girl, she could be described as a child prodigy, perhaps a 'minor celebrity – or rather, a curiosity.'[9] It is clear that her father both encouraged her studies and benefited from the public expressions of her intellectual achievements he organized to take place in their family home. Agnesi was able to use Latin, Greek, and Hebrew while still under the age of ten. She published a discourse in Latin in which she defended the idea of education for women – when she was nine. But it was mathematics and the uses of mathematics to which she dedicated most of her life.

We think of her here because of her work on calculus and on her contribution to the world of mathematics that she so boldly approached and in such a personal and unique way. In 1748 she was the first woman in history to publish a mathematical book, a treatise on calculus. This book, was of course her now famous *Instituzioni analitiche ad uso della gioventù italiana* (*Foundations of Analysis for the Use of Italian Youth*, Milan 1748), her interpretation of Newton's differential calculus.[10] In this book too, she pursued her own study of mathematics and argued for the woman's right to access higher education and the learning of higher mathematics. She dedicated the book to Maria Theresa, the Empress of Austria, and in return was gifted jewels from the Empress. In 1750 Agnesi was appointed as a first female professor to the chair of mathematics. This position was based at the University of Bologna and the appointment came from Pope Benedict XIV no less. The Pope was a student of mathematics in his youth, and admired the book that Agnesi wrote and published two years previously. The two, Pope and Agnesi, corresponded on mathematics.

It is in this book where we meet with the Witch of Agnesi. This is now the name given to a curve Agnesi called versiera, referring to the 'turn' in the curve, or a curve that turns. By chance, Agnesi's book came to be translated into English at the very beginning of the nineteenth century (1801), as it was seen by a group of English mathematicians to be an interesting and original contribution to the study of calculus and algebra in a Newtonian tradition. Coming from the continent, this was no small feat. The translation work befell upon John Colson[11] who made what is now known as a classical mistake in at least the point of translating the name of the curve 'la versiera': He took it to be a misprint, added a few letters, which made it 'l'avversiera', and pronto, we have the birth of the most famous witch in the history of mathematics, the curve now commonly called *Witch of Agnesi*.

The curve was actually first mentioned, but not named, by Fermat at the beginning of the eighteenth century and was constructed and named by the Italian mathematician Guido Grandi in 1718.[12] So the curve had already been in existence for thirty years before being described in detail

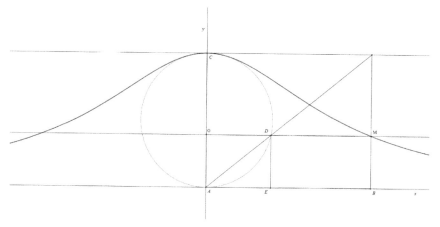

FIGURE 5.3 The 'Witch of Agnesi' is the red line in the diagram. It is, again, a kinetic curve: a curve described by a controled movement.

in Agnesi's work, and then in 1801 it became a world famous curve with a universal name, a witch.

AND, BY A TURN IN THE STORYLINE…

Agnesi was always a highly religious person, and corresponded as we saw even with the Pope. But her religiosity was not devoid of mathematics, nor of the need for women to study mathematics, a cause she pursued all her life. She believed that through the study of mathematics, the mind would learn of certainty and rise above the physical into the world of abstract truths, so beautifully evident in geometry. In her later years she became a full-time theologian and a director of the hospice for women, the Hospice Trivulzio for the Blue Nuns in Milan, where she eventually joined the sisterhood. And what about Algarotti?

Algarotti's book was an easy read for the ladies – presenting ladies as creatures who can, after all, learn about how the world works through discussions on science (but not mathematics) as they are carefully led around the lake by a gentile chevalier. One wishes that at some point such a gentile chevalier would take them around such a lake several times, and gently also push them into the lake so that the pomposity and pretension of the status of women would wash away through indignity of such a scene! Or otherwise – should the marchioness have gently, as if imaging the curve with such a bewitching turn as Agnesi's own, have pushed the chevalier into the lake to end the patronizing treatment she suffered in being taught science in a different way to that of others only because of her sex?

We must not be distracted with such thoughts. The fact is that there were women at the same time Algarotti wrote his book, like Bassi and Agnesi, doing real science and mathematics. They were known and celebrated, and even supported at the very top of their society to partake fully in the intellectual life of their country. Another such Newtonian 'lady' was certainly Émilie du Châtelet, a long-term mistress and companion of Voltaire, whose translation of Newton's *Principia* is still considered the standard translation of the work into the French language. And, to be fair to Algarotti, he was somehow connected to most of the Newtonian ladies who lived in his time, and his book, although seemingly patronizing, an archetypal 'mansplaining' book on popular science, was actually a celebratory ode to women, and to the study of what was then a new and revolutionary approach to understanding the world through Newtonian science. Angesi is, in this context, safely a queen and her study of *la versiera* has ensured that the history of mathematics has a witch of its own.

NOTES

1 *Philosophiae Naturalis Principia Mathematica*, or in short, *Principia* was a major work of Isaac Newton, in three books and written in Latin. The first edition was published in 1687, two more authorised editions were published by Newton in 1713 and 1726. This work states Newton's laws of motion, his law of universal gravitation, and derivation of Kepler's laws of planetary motions.

2 Francesco Algarotti (1712–1764) was a Venetian polymath, essayist, and writer.

3 Madame du Châtelet (1706–1749) was greatly admired by Frederick the Great of Prussia (1740–1786), who re-established the Academy of Sciences in Berlin. They corresponded over years, and it was he that introduce Châtelet to Leibnizian philosophy. In return Châtelet sent him a copy of her translation of Newton.

4 Emperor of Poland, Augustus III (1696–1763) was a King of Poland and Grand Duke of Lithuania, as well as Elector of Saxony. A powerful man, he was also a prominent man of culture, supporting various causes for the promotion of new science and the arts. Johann Sebastian Bach (1685–1750) for example, dedicated his Missa in B minor to him.

5 The tradition we inherited from classical Greek dialogues, and that resembles the dialogue of Meno. Meno, as many a mathematician knows, is the title of one of Plato's dialogues, written in the form of Socratic dialogue. The reason it is a popular read for mathematicians is that some serious mathematics appears in it, describing possibly for the first time, in an explicit way, the concept of incommensurability. Incommensurability means that there are such quantities which cannot have a common measure. Two such would be a side and a diagonal of a square, as an example that appears in this dialogue. See Plato (2009).

6 Francis Yates in her book *The Art of Memory* gives a history of the technique that is used by Algarotti's text, as explained here.

7 Laura Bassi (1711–1778) was the first woman to officially teach at a university in Europe: She was appointed a professor of Anatomy in Bologna, and later was given a Chair in Experimental Physics at the same university.

8 François-Marie Arouet, better known as Voltaire (1694–1778) was a French philosopher, writer, and historian.

9 This is a quote from Massimo Mazzotti's book *The World of Maria Gaetana Agnesi, Mathematician of God*, published by JHUP 2007, p. xi.

10 The second volume was published a year later.

11 John Colson (1680–1760), was an English mathematician and Lucasian Professor of Mathematics at the University of Cambridge.

12 Guido Grandi (1671–1742), Italian monk, theologian, mathematician, and engineer. Guido was born Luigi Francesco Lodovico, but after he took orders he changed the name to Guido. He also studied *versiera* (from Latin vertere, to turn). His book in which this appears is *Relazione delle operazioni fatte circa il padule di Fucecchio*, and was published in Firenze 1718.

CHAPTER 6

June

MATHEMATICS IS LIKE… A TRANSLUCENT OBJECT FLOATING IN SPACE

Mathematics can mean different things to different people at different times. It can mean, as we will see, that through its practice a deity can be approached, and some new things learnt, as well as old things rediscovered. Mathematics makes mathematicians and in June we think of Luca Pacioli, a Franciscan friar, mathematician and mathematical author, who died in June 1517. The famous portrait of him by Jacopo de'Barbari (painted around 1500), tells us not only how he looked but takes us to a scene from his life. There he is standing in front of us, to be seen as if we have just stumbled upon his mathematics lesson. In this chapter we will look onto this scene, as if from another room through a half-opened door, and try and make out what the diagrams, people, and objects in this painting signify. The mathematical object in this painting, shown suspended from the ceiling, is the image of rhombicuboctahedron, an Archimedean solid that was rediscovered by Pacioli.

S TAYING IN ITALY FOR a little longer, we will now meander through the mathematical landscape of an earlier era, the Renaissance. Once there, one would soon stumble upon the famous Renaissance mathematician Luca Pacioli, who died aged 70 in Sansepolcro, Tuscany, the town of his birth. He has been credited by the invention of a double-entry system of bookkeeping, as it first appeared in his *Summa de Arithmetica,*

FIGURE 6.1   The Portrait of Fra Luca Pacioli (1447–1517) with a pupil, painted by Jacopo de'Barbari, c. 1495–1500. Reproduced with permission from the Italian Ministry of Cultural Heritage and Activities – Museo e Real Bosco di Capodimonte.

Geometria, Proportioni et Proportionalita, printed in Venice in 1494. The book summarized mathematics as the title suggests, but it is interesting for another reason. This double-entry system of accounting is very much still in use today and forms the basis of the development of an effective system of accounting, and hence has sometimes even been credited with the influence on the development of economic and political systems.[1] This book, for the first time, contains algebraic text in vernacular Italian, an increasingly important trend that spread across Europe. This meant that the learning was able to spread amongst the population who did not have access to higher learning through Latin and Greek. But let us leave that aside, as the main reason we are interested in Luca is not his development of this practical application or his work on the promotion of learning mathematics, but his work on geometry and his links with other great geometricians of his era like da Vinci, Piero della Francesca, and Dürer.[2]

Let us therefore observe Luca in his surroundings. The famous portrait, painted by Jacopo de'Barbari sometime between 1495–1500 shows

Pacioli in the center with his tools – blackboard, books, compasses and other drawing instruments, and two geometrical solids. He is facing us, with an expression of concentration, and with his eyes set firmly into a distance, looking beyond rather than at us. He is confident in his posture and appears as a man that possesses important knowledge, in some way so important it may be divine. Behind Pacioli is a pupil, a colleague, or a friend – despite the striking resemblance with Albrecht Dürer we can't with certainty say who that is, and there are different interpretations for this figure. Nonetheless, Dürer was heavily interested in the types of geometrical solids that Pacioli has around him in the painting.

SOLIDS NAMED AFTER A GREEK PHILOSOPHER

Here is a fact that you probably already know, but let us play a little game anyway. I'm sure there are some other names that are given to objects ('define what an object is' would probably be what a philosopher would say), but the most famous, regular ones of all mathematical objects in three dimensions, are named after one of the most famous philosophers from ancient Greece.

Let us then look at the mathematical objects appearing on Pacioli's portrait. One we know is a dodecahedron; it is one of the five Platonic or regular polyhedra, known since antiquity – the only five regular solids in three dimensions. So there you have it! This is a regular polyhedron named after Plato, one out of the five, and there are only five *regular* polyhedra in three dimensions.

These solids are regular convex bodies and are a subset of a three-dimensional space – they are each bounded by congruent polygons, meaning their sides are exactly the same in both shape and size for each solid individually. A Platonic solid has the same number of faces and the same number of edges meeting at every vertex: A cube has three squres and three edges meeting in exactly the same way at each of its eight vertices; a dodecahedron has three faces and three edges meeting in the same way at each of its twenty vertices, and so on. And to reiterate, there are only five such solids, tetrahedron, cube or hexahedron, octahedron, dodecahedron, and icosahedron. It is interesting to note also that there are only three regular polygons that make up these solids: the tetrahedron, octahedron, and icosahedron are bounded by equilateral triangles, the cube is bounded by squares, and dodecahedron is bounded by pentagons.

They are named after Plato as he gave us the complete set and described them in his dialogue *Timaeus* (around 360 BCE). But Plato did not claim

FIGURE 6.2   The five regular solids: tetrahedron, octahedron, cube, icosahedron, and dodecahedron.

that he discovered them and there are multiple records to suggest that three solids were known to Pythagoreans[3] – the cube, tetrahedron, and dodecahedron – and all five to Plato's contemporary and collaborator, Theatetus.[4] A persistent and recurrent interest in these rather innocuous objects over many centuries resulted in many developments in mathematics. Proclus, the commentator on both Plato and Euclid, claimed that the understanding of these five regular solids was the primary aim of the greatest book of antiquity on mathematics, the *Elements*. We will see shortly how they influenced the revival of mathematical study in Renaissance Europe.[5]

Plato claimed that the five regular polyhedra are the most beautiful and perfect structures possible, and his dialogue *Timaeus* is written partly to persuade us that this is the reason why they are, in a way, a footprint for the making of our universe. First, he established that there is such a thing as an ideal political arrangement and an ideal society towards which we should aim. Rather than giving a simple answer of how that society should look, he encourages his readers to find the principles of this ideal arrangement in the study of mathematics. He wasn't precisely suggesting that in such a straightforward manner, nor was he offering a couple of perfect models and asking people to choose between what was on offer. Plato indeed goes a little further. He states therefore, in this dialogue between Socrates, Timaeus, Critias and Hermocrates, that this is in fact possible to do, as the whole universe is based on the set of ideal or eternal and perfect structures. There is the world of eternal, perfect, immovable structures, and the world of change, decay, and imperfection. And of the two, it turns out that the second one is for us, the mortals. But we can also strive for the ideal. And here is how Platonic solids gain importance.

The story continues. Before our universe was formed, the four elements, air, earth, fire and water, were mixed. Yet, as order is preferable to chaos, the divine creator brought the order by using perfect structures as a model to create the world. This divine creator, a demiurge (dêmiourgos, as Plato was Greek) therefore imposes a mathematical order on the creation of the universe by modeling it on these perfect solids: Tetrahedron is used for fire, octahedron for air, icosahedron for water, and cube for earth. What about the dodecahedron? It is used as the center of our world, to establish the wholeness of the universe, and some would suggest in centuries to come that this was the principle upon which the divine spark, the love, and life are based.

You may think that this in itself is really beautiful and pause. But the vision of creation does not stop here. It continues with some further mathematics.

The first tree solids have equilateral triangles for faces. These equilateral triangles can be further divided, if one would ever wish to do so, and in the antiquity (and therefore later, when this type of geometry was revived during the Renaissance) this was apparently a thing to do.[6] In fact, that was also part of Plato's dialogue – the perfection of the five regular polyhedra is partially derived from the structure and perfection of the plane surfaces that make them. All plane surfaces can be made from triangles and all triangles can be divided into two smaller ones, right? Well not only that, but of these smaller triangles, and they should be right-angled ones, there are two types: ones which are isosceles and ones which are scalene. Of these perfect triangles that Plato identifies, the first triangle (isosceles) has the sides in the ratio $1:1:\sqrt{2}$.

Now this triangle itself can be seen in the structure of a square (Figure 6.3).

Of course that triangle then, makes the basis for the construction of the sides of a cube, and is important for that polyhedron. What about the other Platonic solids? If you look at the rest of the regular polyhedra in three dimensions, you will see that three of them are made from equilateral triangles, tetrahedron, octahedron, and icosahedron.

If we dissect this equilateral triangle, we can see that the sides of this right-angled triangle are also in an interesting ratio, $1:2:\sqrt{3}$. This is Plato's second perfect triangle, scalene. So we can construct the four out of five regular polyhedra by these two triangles: the isosceles $1:1:\sqrt{2}$, and the scalene $1:2:\sqrt{3}$ (Figure 6.5).

But what about the dodecahedron, the fifth Platonic solid? That is not given in this dialogue and the description of the dodecahedron, the fifth regular polyhedron, is given as the one for which 'remained one construction, the fifth; and the god used it for the whole'.[7]

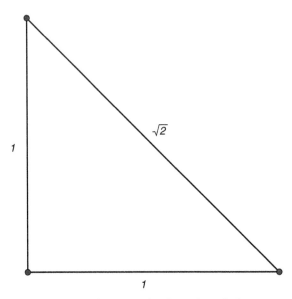

FIGURE 6.3   The isosceles perfect triangle Plato identified.

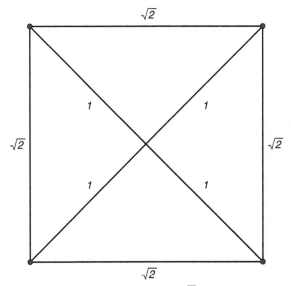

FIGURE 6.4   How to use the triangle $1:1:\sqrt{2}$ to make a square. This is a slightly different arrangement to that usually showing the square of side 1 with its diagonal.

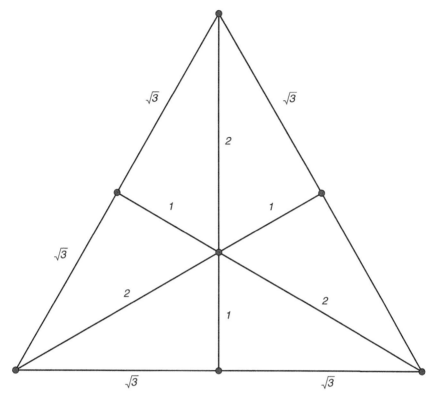

FIGURE 6.5   The isosceles triangles that make an equilateral triangle are used in the construction of tetrahedron, octahedron, and icosahedron.

THE STRUCTURE OF THE FIFTH ELEMENT

One of my favorite movies of all time is the 1997 *Le Cinquième Élément*, or in English, *The Fifth Element*. The film was made in 1997 by Luc Besson and stars Bruce Willis, Gary Oldman, and Milla Jovovich. So why would I mention that here? The plot is actually quite cosmological. Just as in the world of Plato, where the eternal perfection is imitated by the imperfect world we live in, so is this film too, the perfection is maintained by the knowledge of the five elements, which are kept as a kind of template for whenever the chaos, or evil, attacks the world. In order to keep the world going, there must be knowledge and the skill to transform these five elements into being and to get them to defend us against evil. But who can do that? A whole army of well-meaning aliens, or a priest with his clueless and sweet assistant, or an even more clueless but one-time soldier-against-evil, now the very bad taxi driver Korben Dallas, played by Willis. It is the latter who almost

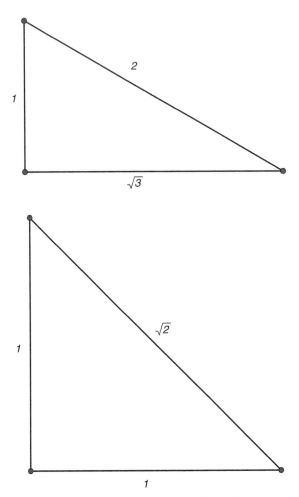

FIGURE 6.6 The two perfect triangles Plato describes that are used for the sides of four regular polyhedra: cube, tetrahedron, octahedron, and icosahedron.

single-handidly saves the world from the evil, the very imperfect and fallible taxi driver Dallas. And he does it by falling in love. Cheesy, yes, but the fifth element, dodecahedron, is the one that has, since antiquity, been identified by that divine spark, which I suppose you can identify with love.

There is a beautiful logic to this film: If the world is structured upon five perfect bodies and evil gets hold of them, it can destroy the world as we will loose the knowledge of how to keep making attempts to recreate perfection. The way Willis saves the world is by recognizing the perfection in his mate, amongst some other things. This is a little convoluted story that leads into some tricky situations, but I will stop here to avoid spoilers.

The point I am trying to make is that it is the fifth element that is really the center stage of this film, and more importantly, it is the fifth element that took center stage in Plato's cosmology.

So let us get back to Pacioli and his portrait. In it, of all possible (five) regular polyhedra, he has one, the dodecahedron – the fifth element. Pacioli, some years after the painting was completed, published a book with the title *De divina proportione*. The most important part of this book, *Compendio de divina proportione* was completed in Milan in 1498, a couple of years before the painting was finished, so it is possible that this portrait was an illustration of his work in that period, and therefore he commissioned it.

There were two other parts to his *De divina proportione*, which were published in 1509: a treatise on architecture and the translation, into Italian, of Piero della Francesca's treatise on regular polyhedra, *De quincy corporibus regularibus*. Pacioli's 'divine ratio', the 'divina proportione' is of course, the division of a line into an 'extreme and mean ratio'. This was not his invention of course. The first description of this division of a line in such a ratio can be found in Euclid's *Elements*, the book VI, Definition 3. It says that,

> *A straight line is said to have been cut in extreme and mean ratio when, as the whole line is to the greater segment, so is the greater to the less.*

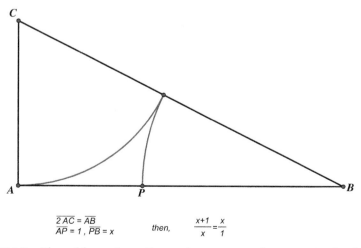

$$\overline{2\,AC} = \overline{AB}$$
$$\overline{AP} = 1\,,\ \overline{PB} = x$$

then,

$$\frac{x+1}{x} = \frac{x}{1}$$

FIGURE 6.7   The golden ratio, or line cut in extreme and mean ratio. Our line AB is cut at point P into golden ratio.

It seems that Pacioli was the first to name this division in this way – to call it 'divine'. There is another popular (current) name for it; 'the golden section'. That came only in the nineteenth century, as 'goldener Schnitt' by Martin Ohm, in 1835.[8]

The divine proportion, the division of a line into an extreme and mean ratio, is needed for the construction of dodecahedron, the solid that is resting on the table in front of Pacioli. In his book on divine ratio, Pacioli gives five explanations to show how this mathematical construction is an aspect of divinity.

In his discussion on the divinity of this ratio, Pacioli's fifth reason is related to dodecahedron:

> *Just as God created celestial virtue (or the quintessence) and by means of this the other four simple bodies, that is the four elements earth, water, air and fire – and through these, everything else in nature – so, according to the venerable Plato in his Timaeus, our sacred ratio gives formal being to the heavens through attributing to them the shape of the dodecahedron (or body formed from twelve pentagons), which, as shall be shown below, cannot be constructed without our ratio.*[9]

This takes us back to the little blackboard that Pacioli is drawing on. It shows a diagram, although we can't make out exactly what it is. There is certainly a triangle there and a circle. Is he beginning to draw a pentagon? If so that is not really a construction from the *Elements*, well not in a way that it is portrayed in the picture. There are obviously some steps that are missing, but that should not be a surprise – he is not after all going to tell everyone how to actually do a construction, instead, he wants this particular picture, a proposition, a theorem, to be picked by those who know Euclid. This is, after all, a precious message about divine proportion and looks a little bit like the diagram in Figure 6.8.[10]

There are other constructions and analyses too that we could work on, which are related to the divine ratio, the pentagon, and the dodecahedron, and we could spend a lot of time playing with them. But let us do one more analysis. If you look at the Figure 6.9, you will see a pentagon with its diagonals. If we denote the diagonals of the pentagon by d, and its sides by s, then we can prove that they are in some kind of sequence.

If we were to continue with constructing propositions from the *Elements* of Euclid, we could go on for a very long time – after all, there is an interconnectedness in the *Elements*, which means it could take us a very long

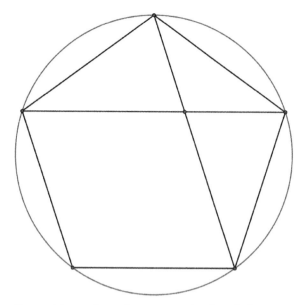

FIGURE 6.8 Proposition 8 from the XIII Book of Euclid's *Elements* says that if in an equilateral and equiangular pentagon straight lines subtend two angles and are taken in order, then they cut one another in extreme and mean ratio, and their greater segments equal the side of the pentagon.

time to cover all possibilities, and perhaps that was the point that the portrait wanted to tell us. The experience of divinity through this practice may be, after all, what Pacioli, according to his own words, found while doing and exploring the Euclidean constructions.[11]

THE TRANSLUCENT OBJECT FLOATING IN SPACE

There is another object in the picture, the translucent solid suspended from the ceiling, half-filled with water. It looks quite magical, but what is it? This object, it seems to me, is the most important one in the picture. We now know what it is and can also name it, thanks to Kepler who did so more than a century after Pacioli's portrait was completed. Now we call it rhombicuboctahedron. It is one of the thirteen Archimedean, semi-regular, polyhedra.

We saw that there are only five regular polyhedra in three dimensions, but there are also some semi-regular polyhedra, and they were first studied by Archimedes. We don't actually have anything from Archimedes that states this, but another commentator on ancient mathematics, Pappus[12] tells us in his *Collection* that Archimedes had discovered thirteen solids

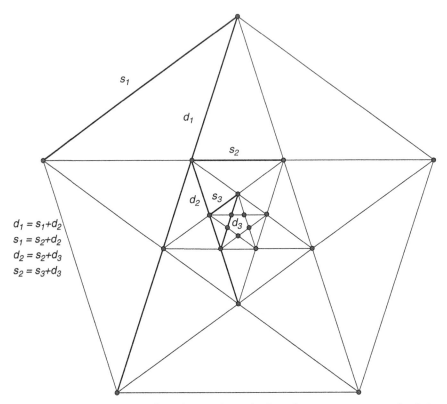

FIGURE 6.9 Pentagon and its diagonals – which make a pentagram – which in turn make a pentagon, etc.

whose faces were regular polygons of more than one kind. That is where the difference lies – the regular polyhedra are made from regular polygons of only one kind, and if you combine some regular polygons you can, as Archimedes discovered, make further thirteen semi-regular solids.

Pappus listed these solids in a descriptive way: for example, a solid of four hexagonal faces and four triangular ones. There was no easy way to imagine these and there were no drawings that survived from the time of Archimedes, although it was known that they were convex and uniform. The defining properties of Archimedean solids are that each face is entirely visible on the outside of the solid, and each vertex is surrounded by regular polygons arranged in the same way.

The thing is, the images, the knowledge, and the ways of construction were lost for centuries, until they were first again studied by the Italian painter and mathematician, Piero della Francesca. In his manuscripts, to

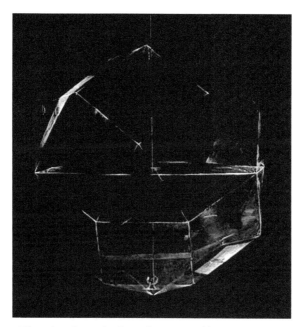

FIGURE 6.10   Rhombicuboctahedron from Pacioli's portrait.

which Pacioli had access (and he has sometimes been accused of plagiarizing Piero)[13] Francesca gives descriptions and some images of the six of the thirteen Archimedean solids. Pacioli goes further. He, in his *De divina proportione*, offers some of the same but also new descriptions, and for the first time since Archimedes, creates the rhombicuboctahedron.

It took, as previously mentioned, another hundred or so years before Kepler would give the first complete set of these solids. He did so in his book *Harmonices mundi*, libri V, which was published in Linz, 1619.[14] And here we can link to another story, and another chapter, dedicated to the last month of the year – and the last in our book, which is dedicated to Kepler. For now, I would recommend you to meditate upon the moment Pacioli discovered the construction and properties of his rhombicuboctahedron: It must have felt like a conversation with a divine force to discover such order and precision, and hence he asked de' Barbari to make the object appear otherworldly, between this and the other world, half-filled with liquid as if containing some ethereal substance. Perhaps this liquid alludes to some further knowledge that is to flow from this object? The sceptic in me, in the meantime, suggests that perhaps, de'Barbari came up with that idea. Who knows?

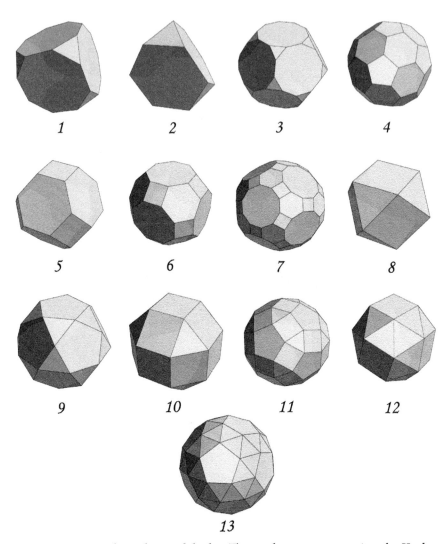

FIGURE 6.11 Archimedean polyhedra. The modern names are given by Kepler: 1 – truncated cube, 2 – truncated tetrahedron, 3 – truncated dodecahedron, 4 – truncated icosahedron, 5 – truncated octahedron, 6 – truncated cuboctahedron, 7 – truncated icosidodecahedron, 8 – cuboctahedron, 9 – icosidodecahedron, 10 – rhombicuboctahedron, 11 – rhombicosidodecahedron, 12 – snub cube, 13 – snub dodecahedron.

NOTES

1 A thesis that the systematic accounting, identified with the double-entry system, played an important part in developing the capitalist system, was first made by Werner Sombart (1924) and continues to be made even now, and in the popular press, see for example Harford (2017).

2 Piero della Francesca (1415–1492), was an Italian painter and also a geometer. He was also a native of Borgo Santo Sepolcro, as was Pacioli. Albrecht Dürer (1471–1528), was a painter and printmaker from Nuremberg. Leonardo da Vinci (1452–1519) was an Italian painter, writer, scientist, and inventor. He did the illustrations for Pacioli's book *Da Divina Proportione* written in 1497 at the court of Ludovico Sforza (1452–1508), Duke of Milan. Sforza was probably better known for his commission of *The Last Supper* by da Vinci, painted around 1490.

3 'Pythagoreans' is generally the name given to the followers of the Greek philosopher Pythagoras (c. 570-490 BCE). They reportedly first lived at Croton in Italy, but later retracted to Greek mainland and settled around Thebes and Philus.

4 For more detailed account see Lloyd's paper on the history of the Platonic solids (Lloyd, 2012). Theaetetus (c.415-369 BCE) the ancient Greek mathematician, was probably the first to establish the group of five regular polyhedra. This conclusion is based on Scholium I of Euclid's Elements XIII. It states that 'the five so-called Platonic figures which however do not belong to Plato, three of the aforesaid five figures being due to the Pythagoreans, namely the cube, the pyramid, and the dodecahedron, while the octahedron and the icosahedron are due to Theaetetus' (Heath, 1936: III, p 438).

5 Proclus Lyceaus (412–485 AD) was sometimes also called the Successor or Diadochos, was a Neoplatonist philosopher, setting out the Neoplatonic system based on his reading of Plato. For us his more important work is his *Commentary* on Euclid's *Elements*. See also Fowler (1999), and Sanders (1990).

6 Philip Sanders (1990) deals with this in detail. See also Moyon (2018).

7 This quote itself is to be found in *Timaeus*, 55c. Translated by Cornford, *Cosmology*, p. 218.

8 An extensive bibliography on the use of these various terms to describe the Definition 3, Book VI, you can see in Sarton (1951) and Wittkower (1960).

9 This quote is from *De divina proportione*, edited by Biggiogero, Fontes bibliothecae Ambrosianae, 31, Milan, 1956, p. 21.

10 The diagram may also be IV.11 (Proposition 11, Book IV) which shows that it is possible to inscribe an equilateral and equiangular pentagon in a given circle – the triangle is first constructed, then from it, the pentagon. It is based on the two other and earlier Propositions. II.11 tells how to cut a line in extreme ratio, and so this original triangle can be made, and IV.10 shows

how to construct a triangle with angles of 36^0-72^0-72^0, so our triangle can be done before completing the construction of a pentagon. But it has been widely accepted that the diagram on the painting refers to XIII.8: If in an equiangular pentagon straight lines subtend two angles [are] taken in order, then they cut one another in extreme and mean ratio, and their greater segments are equal to the side of the pentagon, as in Figure 6.8.

11 All the text for Propositions given here is given in the format from Heath (1908).

12 To gain some perspective on the time which elapsed between Archimedes and his commentator, Archimedes lived c. 287–c. 212 BCE and Pappus some five hundred years later, c. 300–350 AD.

13 The rhombicuboctahedron, the translucent object subtended from the ceiling behind Pacioli, painted (c. 1500) by de'Barbari (c. 1460–1516) was actually for the first time visualized, or drawn and recorded, in this image, as far as we can tell. This is interesting as there have been some charges of Pacioli commiting plagiarism of Piero della Francesca's work. For the full story on this, see Stakhov (2009). The proof that this may not be so was to be found, not in the pudding this time, but in this painting of Pacioli.

14 These appear in the *Elements* Book 2, Proposition 28. The modern names of the solids are those given by Kepler.

CHAPTER 7

July

THE POWER OF (MATHEMATICAL) BOOKS

The last of Shakeshpeare's plays was about a tempest. The main protagonist of The Tempest is the sorcerer Prospero, who was, by numerous accounts, based on John Dee, an English mathematician, astronomer, cryptographer, and a conjurer of spirits. As in life, so in Shakespeare's play, strength and wisdom are gained from learning, and preserved in libraries. Books are to be nurtured, collected, read, and learnt from. To mathematical circles, Dee, who was born on 13th July 1527, is most famous for writing the Mathematical Preface to the first translation of Euclid's Elements into vernacular English, published in 1570. Similar to Prospero, Dee faced many unfortunate events in his life, but as all tempestuous tales, his too has a twist: Via the tempest around the time of the English Revolution and then traveling across the Atlantic, one of the books from Dee's collection found itself in a library in the United States decades after both the death of Dee and of Shakespeare, leaving a lasting legacy of the study of John Dee's work in the new country.

IN SHAKESPHERE'S *THE TEMPEST*, the main protagonist, sorcerer Prospero, conjures up a storm in a game of power between himself and his brother, in order to seek revenge. But to begin from the beginning, Prospero, Duke of Milan, is betrayed and exiled with his daughter to an island. This is no ordinary island as its only inhabitants are magical creatures, and Prospero uses his magic to rule over them. When some dozen years after he is exiled,

FIGURE 7.1 Portrait of John Dee (1527–1608), holding dividers and scroll. Credit: Wellcome Collection. CC BY.

he learns that those enemies from Milan who ousted him were sailing near the island, he creates a storm so strong that it results in the shipwreck of his enemies. He assures his gentle-hearted child that no harm will come to his enemies. Eventually (with much a do in between), through love and compassion (that his daughter inadvertently teaches him as she falls in love with one of the shipwrecked passengers), Prospero and his old enemies suddenly rediscover reason. All animosity is forgiven between them, Prospero's daughter marries the prince, who turns out to be the son of the

King of Milan, and Prospero returns as Duke of Milan to his rightful place. The clear skies are to be seen again, as is usual after a storm.

Whilst the play can be interpreted in different ways, the themes of revenge, forces that exist between people and their positions in society, in addition to a belief that love eventually conquers all, each sit beside a story of mathematics. Let us explore this notion and go behind the scenes and see whether there's any jabberwocky[1] business going on. We now know that the character of Prospero, a lonesome, magic loving, bookish duke was, almost certainly, based on John Dee. Dee was not an ordinary man by any stretch of the imagination. An English mathematician, astronomer, cryptographer, and a book collector, he was also a conjurer of spirits and travelled around Europe to work with and learn from people interested in the similar dark arts as he.

Dee, as previously mentioned, wrote the *Mathematical Preface* to the first translation of *Euclid's Elements* into vernacular English published in 1570.

Euclid's *Elements*, the book we came across in earlier chapters too, is the collection of writings created by Euclid in Alexandria around third century BC. The book was disseminated and used from the time it was written up to the late antiquity, but then disappeared from the syllabus and was rediscovered by the western world mainly through its Arabic translations. The first translation from Arabic was made by an English monk Adelard of Bath,[2] who translated it into Latin in 1120. Then the first printed edition was made in 1482, by Erhard Ratdolt in Venice, based on another thirteenth century copy made by Campanus of Novara. It is necessary to mention this as it demonstrates that centuries passed between different editions of this book; certainly that testifies how well it was regarded by mathematicians over a very long period of time. Why otherwise translate it in so many languages over so many centuries?

The text itself did not change hugely through these translations – it contained the same structure with the same meaning of words. Sometimes there were small differences to be found in the presentation of the text or the understanding, but nothing that would change the mathematical content. This was the most commonly used book from which mathematics was learnt until the nineteenth century. It was as complete and as well structured mathematical syllabus as one could imagine and as such, the book was used as the most important mathematical treatise for many centuries. From novice to advanced student, everyone was learning Euclid's *Elements*.

So if you were a student in the sixteenth century could you buy and learn mathematics by yourself from this amazing repository of mathematical knowledge passed to us from antiquity? The answer is certainly yes. You would need a good deal of money to afford a book like this (none were printed in England up to the late sixteenth century) and you would also need to be able to read either Latin or Greek.

In the sixteenth century though, some realized that such knowledge as contained in the *Elements* can be transferred perhaps more easily if it was given in the vernacular language. Where to find the words to express such superlative knowledge as was contained in books like the *Elements* was another matter. As John Skelton's poem so charmingly stated at the time,

Our language is so rusty
So cankered and so full
Of forwards, and so dull,
That if I wold apply
To write ornately
I wot not where to find
Terms to serve my mind.

John Skelton, *The Boke of Pylyp Sparowe (c. 1504)*, lines 777–783.

To the rescue came Marlowe, Shakespeare, and for mathematics, John Dee with his friend Sir Henry Billingsley.[3]

THE MATHEMATICAL PREFACE

The first book that appeared in English language that taught Euclidean geometry to English readers did not have much to do with Dee. It was written by Robert Recorde and was not a translation of the *Elements* but an adaptation of the first four books from this collection. Recorde wrote the text himself and his book reads like an abridged version of the *Elements*.[4]

Later in the century, a wealthy merchant and haberdasher, later also Lord Mayor of London, Henry Billingsley, undertook the task to translate the full *Elements* into English. His translation appeared in 1570 under the title *The Elements of Geometrie of the Most Ancient Philosopher Euclide of Megara*. As it turns out, he was wrong there, as the correct Euclid that wrote or compiled (we'll come to this story in Chapter 8) the *Elements* was the Euclid of *Alexandria*. There was an Euclid of Megara at the same or similar time and he too was a philosopher, but he was not our Euclid. This appears to be the only major mistake in Billingsley's edition. His *Elements* were otherwise renowned for their accuracy and clarity of language. In later years the translation was credited to John Dee, by the first professor of mathematics at University College London, Augustus DeMorgan, and in the years to come there were other suggestions that the translation was not by Billingsley at all, but by his friend and protégé Augustian friar Whytehead who lived in Billingsley's house.[5] It is a rather curious coincidence that quite a lot of the writing related to the men who are of most interest in this chapter is in some way doubted. Shakespeare's plays are said to have been written by Marlowe, Billingsley's translation of Euclid by Dee or Whytehead, and there are at least three theories about whether Euclid wrote the original *Elements* himself or compiled the writing by others (see Chapter 8, for August). In the case of Billingsley, we can almost certainly say that it was really him that translated Euclid into English as eventually a copy of the Greek text of Theon's Euclid was found in Billingsley's house, full of notes of his comments in his writing.

John Dee wrote the preface to this, the first English edition of the *Elements*. And from this beautiful piece of writing about mathematics, we can see not only what Dee thought of it but learn of the tradition going back as far as Ptolemy, in which mathematics was seen as a discipline to train the mind and through which one could achieve the understanding of the mysteries of the universe. Dee too, like his Italian predecessor about whom we learnt in Chapter 6, describes this power contained in mathematics as being divine:

> *But unto God our Creator, let us all be theknefull: for that, As he of his Goodnes, by his Power, and in his wisdom hath Created all thynges, in Number, Waight, and Measure: So to us, of hys great Mercy, he hath reuelead Meanes, whereby, to atteyne the sufficient and necessary knowledge of the foresaid hys three principall Instrumentes: Which Meanes, I haue abundantly proued unto you, to be the Sciences and Artes Mathematicall.[6]*

BURNING BOOKS

History teaches us how both books and mathematics are often perceived as dangerous knowledge and can be prohibited, burned, or otherwise destroyed in order to stop the spread of such knowledge. Dee had the misfortune to experience this during his lifetime and through his own library. His collection was deemed to be one of the most important libraries in

Renaissance Europe of the time. The inventory of his collection, which Dee compiled before he embarked on a six year trip to Eastern Europe, listed some of the most prominent works on mathematics, mechanics, optics, astronomy, technology, and other subjects. Among his books were also to be found scientific instruments and natural wonders he collected during his lifetime. And there were objects he used when conjuring up spirits and studying the world of the occult – his black mirror made of obsidian (volcanic glass) was brought to Europe from Mexico after Cortes's conquest of the region. We don't know how Dee got hold of such a rarity, but we know that he used it to summon visions of angels and spirits. Dee's work on such occult matters brought him both admiration and, it seems, a good amount of contempt from those who knew him. His perhaps open-hearted demeanor did not help matters. A one time friend persuaded him that for some occult reason or another, they should swap wives; this didn't bring any benefit to Dee and from his notes we can see the turn his life took after that event. He felt betrayed, tricked and this event possibly broke his sprit and belief in human kind.

And what about his dearest possession, his library? Have these excursions into the world of the occult honed Dee's sense of intuition? Of course we can't tell even if he believed that such experiments could have an effect on human abilities. But we do know that Dee had a premonition about his own library being dismantled, as he recorded two dreams he had of this (before the actual event), one of which he described as a nightmare in which death and the destruction of his books were equated with each other:

> *I dremed that I was dead, and after my bowels were taken out I walked and talked with diverse, and among other the Lord Thresorer who was come to my house to burn my bokes... (Dee, 1582).*[7]

Dee's library, containing some three thousand books and about a thousand manuscripts and objects when he left it, according to his inventory, was raided and pretty much destroyed it seems, as soon as he departed his home in Mortlake, London in 1583 to travel around continental Europe. His books, scientific and occult instruments, and curious objects were stolen, most probably by his students, friends, and neighbors.

AND THEN, ACROSS THE POND

As with all tempestuous tales, so this too has a twist: Via the tempest of the English Revolution, and then traveling across the Atlantic, one of the books from Dee's collection found itself in the library of the son of the founding

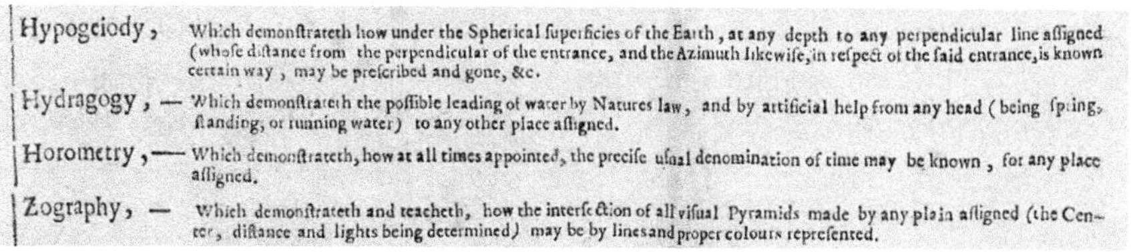

Hypogeiody ,	Which demonstrateth how under the Spherical superficies of the Earth , at any depth to any perpendicular line assigned (whose distance from the perpendicular of the entrance, and the Azimuth likewise, in respect of the said entrance, is known certain way , may be prescribed and gone, &c.
Hydragogy , —	Which demonstrateth the possible leading of water by Natures law , and by artificial help from any head (being spring, standing, or running water) to any other place assigned.
Horometry , —	Which demonstrateth, how at all times appointed, the precise usual denomination of time may be known , for any place assigned.
Zography , —	Which demonstrateth and teacheth, how the intersection of all visual Pyramids made by any plain assigned (the Center, distance and lights being determined) may be by lines and proper colours represented.

FIGURE 7.2 Part of Dee's classification of the diagram showing his classification of mathematical sciences, with the emphasis on the less well known and today considered more to be 'jabberwocky'.

governor of Massachusetts Bay Colony.[8] The book was the sixteenth century copy of the second edition of Apollonius of Perga's mathematics. On the page facing the frontispiece to Apollonius's book, Dee for the first time sketched a draft structure diagram that showed how he understood the interconnectedness of mathematical sciences. This perhaps is the first draft of Dee's classification of mathematical arts and sciences, and it resulted eventually in the diagram he produced in his famous *Mathematical Preface*.

In this diagram we find all kinds of mathematical studies, apart from arithmetic, geometry, music, or astronomy: There are branches that sound like some 'jabberwocky disciplines' (Clucas 2006, p. 6), for example Trochilike – which in Dee's words "demonstrateth the properties of all Circular motions: Simple and Compound" or 'Helicosophie' straight after it which is concerned by the study of "all Spiral lines: in Plaine, on Cylinder, Cone, Sphaere, Conoid, and Sphaeroid: and their properties".[9] Today we have a very different classification of mathematical sciences, but this first complex diagram plotting the different branches of mathematical study was then, and still is, groundbreaking. It has also allowed generations of learners to look at different aspects of abstract and practical knowledge and skills that mathematics deals with in a structured and clear way. While today we have for example a pure and applied mathematics as two broad terms within which a myriad of mathematical fields are contained, Dee's vision of mathematics went further and included an awareness and even a level of the study of the supernatural. Mathematics, for Dee, was a field that connected the supernatural with the natural.

> *Things supernatural, are immaterial, simple, indivisible, incorruptible and unchangeable. Things natural, are material, compounded, divisible, corruptible and changeable... In things natural, probability and conjecture hath place; but in things supernatural, chief demonstration and most sure science is to be had. By which properties*

and comparisons of these two, more easily may be described the
state, condition, nature and property of those things which we before
termed of a third being; which, by a peculiar name also, are called
things mathematical... (Dee, 1582)

Most importantly, beyond Dee's diagram of the classification of mathematical arts, his *Groundplat*, there is his simple unifying idea that mathematics is a "strange participation between things supernatural, immortal, intellectual, simple and indivisible: and things natural, mortal, sensible, compounded and divisible."[10]

Dee's project was not a little one. It was to understand the world and to be able to articulate the connections between the two worlds, the natural and supernatural, through the study of mathematics. Our perception of what he called 'supernatural', and certainly the methods he employed to understand it, had changed immensely, and one would most probably laugh at a mathematician today trying to conjure spirits by a black mirror. But his proposition is still valid in the sense that mathematicians still search to understand the unseen, the unknown, and discover mathematical principles of the hidden structures, patterns, and see how they can describe the world. His preface is dedicated to the "sincere lovers of truth, and constant students" – and to them he wishes as a reward a "grace from heaven, and most prosperous success in all their honest attempts and exercises".[11] And apart from this, what would be the motivation for you to study mathematics, according to Dee?

Many other arts also there are which beautify the mind of man: but
of all other none do more garnish and beautify it than those arts
which are called Mathematicall.[12]

NOTES

1 This name calling of certain mathematical sciences as 'jabberwocky' is due to Westman et al., and was cited in Clucas (2006, p.6). See also Lawrence (2011).
2 See Burnett (1987); (1999).
3 Sir Henry Billingsley (c.1533–1606) was an English merchant and a Mayor of London. He was the first translator of Euclid's *Elements* into English language.
4 Robert Recorde (1512–1558) was a Welsh mathematician and physician. He is credited by the invention of the '=' sign, and the introduction of the '+' sign. Recorde published a number of books on mathematics and medicine in English language, one of which is his book on geometry, *The Pathway to Knowledge*, published in 1551.

5 Augustus De Morgan (1806–1871) was the first professor of mathematics at the University College London (originally called London University) founded in 1826. For further reading on the authorship of the translation of Euclid's *Elements*, see Halsted (1878), Ball (1889) and Archibald (1950).

6 John Dee, *Mathematical Preface*, p. 33.

7 See also Lawrence (2011).

8 English Revolution, also known as the English Civil War, took place between 1642 and 1660. This one of Dee's books was bought by John Winthrop the Younger, in the year when he was about to leave for America to join his father, the first governor of the Massachusetts Bay Colony, in 1631. Through Winthrop Jr., Dee's work spread to Puritan New England (see Woodward, 2011; Calis, et al., 2018).

9 Both quotes appear on Dee's (1570) Groundplat, part of his *Mathematical Preface*.

10 Dee (1570).

11 Dee (1570).

12 Dee (1570).

CHAPTER 8

August

REVOLUTIONARY MATHEMATICS

As you travel through mathematical landscapes, you will come across some dangerous bends. Beware of them, be prepared and pause to reconsider what you know. Be prepared for a change of scenery and remember that even revolutions are known to have taken place in mathematics. With revolutions in mind, we turn to the twentieth century to the place where the French revolutionary mathematics was first taught, the École Normale in Paris, after the French Revolution of the eighteenth century. Some centuries later, a group of young mathematicians conceived the new way mathematics was to be presented and taught in the twentieth century. We look at the mathematics that was made in the build up to World War II by this young group of mathematicians, French and German, from which emerged one of the most productive mathematicians of our times, the ever-present Nicolas Bourbaki.

O N 30ᵀᴴ AUGUST 1952, a young mathematician, Nicolas Bourbaki, was given a home at the École Normale in Paris. Although only eighteen, he was at the time already a published author. Bourbaki became probably the most famous imaginary mathematician of the modern era.

In early 2015, I spent a month in Nancy, a beautiful Art Nouveau city in the East of France, at the University of Lorraine's Henri Poincaré Archive. I did some work there for their Masters Programme in the history of mathematics, but I was also there to exchange ideas with French colleagues and

FIGURE 8.1   Nicolas Bourbaki Congress in 1938. Simone Weil (1909–1943), a sister of André, Charles Pisot (1910–1984), André Weil (1906–1988) standing behind Charles, Jean Dieudonné (1906–1992), Claude Chabauty (1910–1990), Charles Ehresmann (1905–1979), Jean Delsarte (1903–1968).

soak up their valuable and particular take on the understanding of mathematics and mathematical cultures. A beautiful library with open shelves, adorned the institute, and I quite literally climbed its shelves in disbelief to see how many different, rare, and beautiful books on mathematics had been collected. And a whole shelf on mathematics and aesthetics certainly made me feel at home.

Whilst there, I was also reminded of the cultural and mathematical heritage of the University of Lorraine and Alsace-Lorraine region, and so the following story about the French mathematician Bourbaki is closely related to this region and the collaboration of a group of French and German mathematicians that arose before and endured the Second World War.[1]

THE REAL BIRTHDAY OF IMAGINARY BOURBAKI

André Weil, Auguste Delsarte, and Henri Paul Cartan, were mathematics students at the École Normale in Paris, one of the grand schools of France in 1935. During the preceding year they were considering publishing textbooks together that they could all use in their teaching. In 1935 though, they saw a notice that a mysterious mathematician was to deliver a lecture on famous theorems. This mathematician was very glamorously dressed and presented a series of completely fictitious theorems, all named after French generals (was this a play on Napoleon's theorem itself?)[2] and finished with the one named after General Bourbaki. This last theorem, in particular, built on all the previous theorems given by the speaker. Weil described this last theorem in his autobiographical account as "taking off from a modicum of classical function theory, rose by imperceptible degrees to the most extravagant heights" which left the "audience speechless with amazement".[3] No one, as far as I can tell, could now describe this theorem. But there is some record of the 'famous' mathematician – it was in fact a student, Raoul Husson, who dressed in extravagant clothes, wore a false beard, and spoke in an invented foreign accent. Husson later became a famous phonologist, so his accent must have been convincing.

Over the coming years, the three students, Weil, Cartan, and Delsarte, remembered this occasion and conspired to create a fictitious mathematician named after this theorem. In 1935, as the Nazi's rose to power and their politics threatened the stability of Europe, the young group which had been growing and included both French and German mathematicians, decided to publish under an assumed name of Bourbaki, in Comptes-Rendus, one of the seven publications of the French Academy of Sciences. The question arose what first name they would adopt for General Bourbaki and how to present this new mathematician: Eventually Weil's wife Evelyne gave him a name of Nicolas and thereby became Bourbaki's godmother. Nicolas Bourbaki was born and still lives despite old age in the École Normale in Paris. He is truly a testimony to non-hierarchical communal survival that is possible in the mathematical world.

Bourbaki, who later collectively included Jean Dieudonné, Laurent Schwartz, Jean-Pierre Serre, Armand Borel, and Alexandre Grothendieck, to name a few famous mathematicians who were part of the group, wrote a series of books under the name *Éléments de Mathématiques* (*Elements of Mathematics*), which aimed to be a self-contained summary of the core areas and ideas of modern mathematics. Bourbaki's style is highly personal, well received, and these books can be found in the mathematics sections of many university libraries around the world.

THE MODERN ELEMENTS

Does this mention of the *Elements* remind you of someone else? Of course, Euclid of Alexandria, who could be compared with Bourbaki and his own modern *Elements*. The French mathematician and historian Itard, most probably inspired by Bourbaki's work and existence, came up as a consequence with a new theory. In his 1959 book, *Mathématiques et Mathématiciens* he proposed that the ancient Euclid may also have been a group of mathematicians and not necessarily one person. He suggested three hypotheses:

1. Euclid was a historical character, known as Euclid of Alexandria, who was born around 325 BC and died around 265 BC, and was the mathematician who wrote the *Elements* and other books.
2. Euclid was the leader of a team of mathematicians working in Alexandria around the same time, 300 BC, and the team together wrote the *Elements* and other works attributed to Euclid. The group may have even continued to write after their leader's death.
3. Or Euclid, like Bourbaki, is not at all a historical character. Perhaps, for some reason, a group of mathematicians wanted to write a collection of all mathematics known to them at the time, and did not want to identify themselves, and so invented a pseudonym.

This last case has a particular charm and attraction for us, but whether it is historically accurate is anyone's guess. All versions however, suggested by Itard, may be considered to be equally probable, as all we know about Euclid is what we have learnt from secondary sources.[4]

Perhaps, on the other hand, the proposition by Itard to consider these options was particularly inspired by the difference of French to other Western European mathematical traditions. I refer of course, to the change that the French Revolution brought to the learning of mathematics. During the French Revolution, the mathematics curriculum in France was drastically changed so that it could indeed be called revolutionary. The old institutions of France were closed and replaced with new, such as the École

Polytechnique in Paris. This school, which opened its doors to students in December 1794, had the aim to promote new mathematics and science[5] and attracted the most famous mathematicians of the time, such as Lagrange, Monge, Lacroix, and Laplace. They wrote books, albeit under their own names, that for centuries since have inspired others to think in new and different ways – revolutionary books for revolutionary institutions. This revolutionary group of mathematicians, similar to Bourbaki or Euclid, if we accept the third proposition of Itard, offered role models of non-conformity and freedom of mind and spirit that, at the time when it peaked, gave us mathematics that is timeless and transcends cultures.

THE ETERNAL APPEAL OF PYTHAGORAS

It would be then interesting to discuss in a little more detail one mathematician from the Bourbaki group, and one of its most controversial members.

During my stay at Nancy, I once went with colleagues to the university canteen for lunch, when someone mentioned Grothendieck. My ears perked, as the story was gripping, and I knew that the University of Lorraine was the descendent institution of Grothendieck's *alma mater*. A number of anecdotes were told during that lunch, about him trying to establish a group of pure research mathematicians who would be prepared to live a reclusive life, something alike Pythagoreans, with Grothentieck as a modern-day Pythagoras at its center.

Alexandre Grothendieck (1928–2014), one of the Bourbaki members, is probably best remembered in mathematics for his invention of the 'theory of schemes' and 'motives'. He developed this in his *Élements de géometrie algébrique*, which he published in 1960 whilst based at the Institut des Hautes Études Scientifiques, France (IHÉS). In this book he connected algebraic geometry, commutative algebra, and number theory. A few years later, Grothendieck wrote to his friend Jean-Pierre Serre and introduced the notion of a 'motif'. This all sounds pretty exciting and also pretty general – schemes and motives? In fact it is a highly abstract set of mathematical concepts. It would probably be unwise to attempt to explain what they actually mean, as we would need to talk about cohomology, base fields, and consider something called the category of smooth projective varieties.

However, this is not the reason why Grothendieck is mentioned here. Whilst he was, undoubtedly, a mathematical genius and his mathematical work and ideas, although highly abstract, were highly valued, he is at least as well-known for his radical and political activism. He was an unreserved pacifist, which is perhaps not surprising having experienced a tumultuous childhood and adolescence. Grothendieck's father had Jewish background and both of his parents were active anarchists who partook in the Spanish Civil War, during which time they left Alexander in the care of a friend. Later on, his mother survived an internment camp during the Second World War and his father was sent to Drancy and from there to Auschwitz where he died in 1942. During that time, Alexander himself lived in a village in Haute-Loire region, occasionally seeking refuge in the surrounding woods when Nazis visited and raided villages in the region.

By all accounts, he grew to be a highly intense and passionate man. This extraordinary intensity of all of his outputs and activities give us a picture of a person of great intelligence, integrity, and individualism.

So how come in his later years, Grothendieck became a recluse and a mystic? Why and how mathematicians such as him, and of course also Pythagoras himself, become passionately involved with spiritual matters is something that has puzzled me for a long time.[6] To find the answer to this question, some of his non-mathematical writings can be a good starting point. I read excerpts from his *Le Clef des Songes*, which is both wonderfully obscure and enlightening at the same time. This 'key to dreams' is Grothentieck's exposition of his spiritual experience and looks into the connections between physical and psychic realities. At the base of this experience lay his deep belief that dreams are given to us, sent by the *le Rêveur* (the Dreamer), the ever-present force of some kind. In this manuscript he develops this theory further to include a general epistemology of a kind: He talks about how there are different types of knowledge, the external and the internal, which inspires, teaches, and leads each individual throughout their life. He mocks those who, such as Gödel, try to prove or disprove the existence of God.[7] The only such proof, Grothendieck believed, is that given by revelation and experience, quite an unusual thought for an utterly modern mathematician.

And here again, we meet with an image of a stereotypically strange and unworldly mathematician – a vegetarian, an anarchist, a pacifist, a mystic, and recluse. This is, I am sure you already know, not an unusual image of mathematicians, and is often reinforced by the portrayal of mathematicians in movies or popular culture. Think only of John Nash in the film *A Beautiful Mind*, the news surrounding Grigori Perelman's well publicized refusal to accept the one million dollar Millennium Prize, or even Galois's dueling that led to his early death. We have at least three real-life examples, so mathematicians must therefore really be strange creatures?[8] Certainly this perception is not necessarily bad in romanticizing mathematics or mathematicians,[9] but for the sake of accuracy and understanding of what is it that mathematicians actually do, perhaps there is something else that is at play here.

DANGEROUS BENDS

This is a very appropriate place to go on another side-road with Bourbaki and look for the symbol he/they invented for reading, learning, and generally for having a particular type of attitude towards mathematics. The study of mathematics should also teach its students that sometimes we have to be careful in case we get taken to a place where we make mistakes. In other words, certain passages are not to be taken lightly, we need to slow down before such a dangerous bend in a mathematical road, stop and consider, before proceeding. This is almost word for word a description that explains the Bourbaki 'dangerous bend' or 'dangerous, caution' sign, based of course on the usual motorway signs.[10]

Should we apply this to our 'reading' of life, humans, and situations we encounter as we pass through life? Pause, consider, before proceeding. At first sight we may see Grothendieck as an unworldly and strange person, but not because or despite his mathematical genius. To look at this a little more, we should also consider his driving force. In several places in his opus, he makes this quite clear: All he wanted was to discover the truth, whatever that may be, and whilst doing so, retain the child-like quality of seeing something for the first time as children do:

> the child … has no fear of being once again wrong, of looking like an idiot, of not being serious, of not doing things like everyone else.[11]

FIGURE 8.2   A dangerous bend road sign.

Because,

If you pay enough attention, it [the truth] becomes clear – mathematics in large doses expands [to explain everything].[12]

BOURBAKI'S THEOREM – EN FIN

Finally, I must say that I have not yet given up the search for that original Bourbaki theorem, which, after the talk given at the École Polytechnique, inspired the first of the Bourbaki group to give birth to Nicolas Bourbaki. I have an inkling that it is in some way related to the idea that a friendship is more important than wars and inspired a group of young mathematicians to work together and form life-long friendships, despite and in face of increasingly hostile and divided societies of Europe at the time. Joined together to collaborate and work despite such divisions, they were not even interested in claiming individual credit for anything they did and published together under an assumed name. We could summarise that through his life, Bourbaki proved that collaboration, friendship, and communal benefit can survive personal difference, wars, and ideologies, and produce mathematics that is indeed, revolutionary.

NOTES

1 The story is given in much detail by one of the founding members of this group, André Weil (1906–1998), in his autobiography *The Apprenticeship of a Mathematician* (Weil, 1991).

2 Napoleon's Theorem states that, if on each side of any triangle, equilateral triangles are added, the lines connecting their centers will form an equilateral triangle.

3 Weil (1991, p. 100).

4 There are different traditions that speak of Euclid, but none from his contemporaries. See Heath (1908).

5 École Polytechnique was first named École Centrale des Travaux Publics in December 1794, but changed its name to its current name in September 1795. See Grattan-Guinness (2005).

6 One of the results of this interest is the book I co-edited with Mark McCartney, see Lawrence and McCartney (2015).

7 Kurt Gödel (1906–1978) was a Czech philosopher and mathematician. Most famous probably for his Incompleteness Theorems, I mention him here because of his ontological proof for the existence of God. See Anderson (2015).

8 The film *A Beautiful Mind* is a 2001 biographical drama focused mainly on John Nash, the Nobel Laureate in Economics, who made a lasting contribution to mathematics in game theory.

9 Amir Alexander for example wrote about this in his *Duel at Dawn*, see Alexander (2011).

10 An example of what the sign is meant to signify, appears in several of Bourbaki's textbooks, and here is the quote from *Théorie des ensembles*, p. I:8.

> *Certains passages sont destinés à prémunir le lecteur contre des erreurs graves, où il risquerait de tomber ; ces passages sont signalés en marge par le signe « tournant dangereux » (Some passages are designed to forewarn the reader against serious errors, where he risks falling; these passages are indicated in the margin with the sign "dangerous bend").*

11 Grothendieck wrote this in his recollections, 1985, *Récoltes et Semailles*, p. 2.

12 The original French says:

> *Pour peu qu'on a y prête attention, elle [cette vérité] crève les yeux – les maths à grosses doses épaissit. Grothentieck (1985: p. 4).*

CHAPTER 9

September

THE MATHEMATICAL TRAVELER IS AT HOME EVERYWHERE

Mathematics can be a way of travel and a way of survival. On our journey through mathematical landscapes, we meet another twentieth century mathematician, a man who knew everyone and whom everyone knew (in mathematical circles), Paul Erdős. Erdős was truly a global citizen, a man of very few possessions but with many friends around the world. He relied on his friends for intellectual stimulation, collaboration, company, and shelter, and they were in turn enriched by his visits. Paul Erdős, one of the most prolific mathematicians of all time, died on the twentieth of September 1996.

A FRIEND WAS BOASTING TO being in her prime lately, that is, a prime number denoting her age this year. By this definition, everyone is in their prime every once in a while. And then, once one gets into later years, they may put letters behind their name, like Erdős did. He was Paul Erdős P.G.O.M. (Poor Great Old Man), which became L.D. (Living Dead) at the age of 60, A.D. (Archeological Discovery) at the age of 65, L.D. but with different meaning (Legally Dead) at 70, and C.D. (Counts Dead) at 75. Erdős died aged 83, which is a prime number too.

And then there is such a thing called the Erdős number. However, we do not all have an Erdős number! Mathematicians have been identifying themselves by their collaboration with the famous Hungarian mathematician since the paper "And what is your Erdős number?" was published

FIGURE 9.1   Paul Erdős (1913–1996), in 1985, talking mathematics with Terence Tao (1975–), who became a Field Medalist in 2006. Photo credit: either Billy or Grace Tao (parents of Tao).

in American Mathematical Monthly in 1969.[1] How does that work and what is it?

THE NUMBER THAT BEARS YOUR NAME

Erdős was the only surviving child of his parents, both Jewish mathematics teachers. He developed an early ability for mathematics and entered the Budapest University at seventeen. By the time he was 21, he had completed his doctorate there. He first left Hungary to become a guest lecturer in Manchester in 1934 and in 1938 was given a scholarship position at Princeton. From this relatively early time in his life, Erdős became an incessant traveler. Erdős had half of his family perish during the Holocaust – his father, two uncles, two aunts all died, and his mother survived in hiding. He was at a safe distance away from Hungary, but unable to settle. This continued after the war had long ended. In this life of permanent turmoil, the only anchor and safe harbor seems to have been those which he found when visiting his mathematical friends with whom he endlessly discussed and wrote mathematics. Erdős attained more than a dozen honorary doctorates throughout the world and is credited with writing or co-authoring 1475 academic papers with more than 500 collaborators.

So important his network of collaborators became for the development of the twentieth century mathematics that the Erdős number progressed into a project that gained high kudos across the mathematical community worldwide. To get an Erdős number is a fairly difficult task. The American Mathematical Association (AMS) calculator uses the data from *Mathematical Reviews* which includes – and this is a qualifying phrase – the criteria for which publications qualify (obviously from the large set of mathematical journals). The Erdős Number project is based at the Oakland University and the administrators are quite strict about the criteria for some collaborative link to be deemed important enough to lead to Erdős himself, and for the Erdős number to be safely written next to someone's name.[2] Erdős of course has the assigned number 0 – the network begins from himself, and his immediate collaborators can claim Erdős number 1, their collaborators 2, and so it continues. So whilst he worked with about 500 (some say 511 exactly) mathematicians during his lifetime, through this structured network based on Erdős number, the number of mathematicians somehow linked to him reaches about 200,000.

On his permanent journey to create more mathematics, it has been reported that Erdős always identified his next collaborator whilst still living with the current one, traveled to them (that being organized by his previous host), and knocked on their door with the greeting: "My brain is open", staying there as long as necessary to complete a paper, or sometimes a few. During his stay with his mathematical hosts, he let them organize his life, which was not that difficult. Erdős was a man of a very few needs and even less possessions – all his worldly possessions could fit into one suitcase. He drank abundant cups of coffee and his colleagues generalized this case to a particular theorem:

A mathematician is a machine for turning coffee into theorems.[3]

Once Erdős and his host would finish the papers they worked on, Erdős would expect his host to take him to the next destination, a next identified collaborator, and a next Erdős No.1.

Erdős was an incredibly generous man, and all his earnings were principally given to various worthy causes. He offered financial prizes for the solutions of unresolved problems in mathematics. Whilst there are some much more financially viable prizes to be hunted by young mathematicians (see for example Clay Institute's prizes), Erdős' problems are in no way trivial and can certainly put the mathematical puzzle solver on the map of Erdősian mathematical network.[4] There is no official list of Erdős problems and after his death, his friend and collaborator Ronald Graham became an informal administrator of the award system.[5] To give you a flavor of what you can work on, here is an example. If you prove Collatz conjecture you can claim five hundred dollars. The conjecture goes like this.

Start from any positive integer (whole number) n, then construct a number sequence (collection of numbers written in a definite order) such that if the previous number is even, the next term must be one half the previous term. If the previous term is odd, the next term is three times the previous term plus 1. The conjecture/hypothesis of Collatz says that no matter what n we begin with, the sequence will always reach 1.[6] For example, start with $n=6$.

Then we have a sequence

6, 3, 10, 5, 16, 8, 4, 2, 1.

You can try yourself! But be careful. I won't help you – there are some numbers that will give you very, *very* long Collatz sequences.

CAN I IMAGINE AN EVEN BIGGER NUMBER?

Before I get to the type of mathematics Erdős did, I'll make a little diversion into talking about some big numbers. With Collatz sequence we saw how we always end with 1 – but what if you thought of imagining as big a number as you possibly can, as far away from 1 as is possible?

This type of question is not really modern. Since antiquity people have wondered how to describe big numbers: sure you can think of a very big number, and an even bigger one, but that's not really enough. Do an experiment with your friend. Say you want to compete with each other by imagining the biggest number you possibly can: How can you be sure that your number is bigger than your friend's? Go even further and say that you settled your number is bigger. But your friend can say that they added 1 to it, so surely their number is even bigger now?

You actually need some kind of structure that would define a number in order to say how big it is. And that structure can literally be a heap of sand.

BIG NUMBERS IN ANTIQUITY

Whether you plan to drive down the Desert Highway in Wadi Rum in Jordan or stroll down some sandy beach of Sicily, counting grains of sand can provide you with an opportunity to think about such big numbers (see Figure 9.2). Just like you, so Archimedes, some time in the third century BCE, could have wandered around some sandy beach of his native Sicily (and has numerous squares and streets named after him there; an example

FIGURE 9.2   A heap of sand in Wadi Rum, in Jordan. Photo credit: author.

FIGURE 9.3   Archimedes Square in Syracusa, Sicily. Photo credit: author.

is shown in Figure 9.3), toying with the idea of how many grains of sand there are in the world. As a consequence, he wrote the *Sand Reckoner* (Ψαμμίτης) trying to produce an estimate of the number of grains of sand that could be counted in the whole of the universe.

Judging by his *Sand Reckoner*, Archimedes seems to have been quite irritated by phrases and beliefs that grains of sand cannot be numbered. We can mention two such examples: When wanting to say that something is innumerable, the Greek poet Pindar wrote that 'sand escapes counting,' and another Greek writer Aristophanes described an immeasurable pain by comparing it to sand, 'my pains? Far more than all the grains of sand'.[7] So like all good mathematicians, Archimedes set himself a task to question this common belief and really think about a seemingly simple question.

In the pursuit of finding this very big number, Archimedes had to establish some kind of system: He had to define the shape, borders, and therefore the actual size of the universe, just for the beginning. Then he had to estimate the number of grains of sand that can be contained in it, name this large number, and therefore work with numbers larger than any that had been worked on before him. So how did he do it?

Archimedes first deals with a myriad, which is 10,000. He expands this easily into the 'myriad of myriads', i.e., 10,000 x 10,000, which is 10^8. He continues but stops to give explicit names to all large numbers, and constructs instead a system of large numbers and then expresses the ranges of numbers.[8] The numbers of the 'first order' are then numbers from 1 to 10^8 (myriad of myriads). The second-order numbers are built upon this number, and are of course $10^8 \times 10^8 = (10^8)^2 = 10^{16}$. Third-order numbers are then the numbers from 10^{16} to 10^{24} and so on. It is worth noting that imagining such large numbers must have been pretty hard as the Ancient Greek number system was based on the 24-letter alphabet and had not had the expressions for large numbers such as a myriad until Archimedes. There's of course a little more to this structure, but for our purposes this is enough to get us started with comparing it to our structures of large numbers. Archimedes came to the conclusion that the universe contains about 10^{63} grains of sand.

Our current system of naming large numbers dates back to the fifteenth century French mathematician Nicolas Chuquet, who came up with the names for billion, trillion, and so on until nonillion.[9] But what about numbers after that? Ever since Archimedes set us on this path, it seems the humanity has been thinking of larger and larger numbers, trying to define them in some way.

ERDŐS AND HIS MATHEMATICAL FRIENDS' NUMBERS

With so many papers he wrote and collaborated on, it is pretty obvious that Erdős was interested in a wide range of mathematical fields. He worked in discrete mathematics (the study of discrete structures, such as for example graphs), graph theory, number theory, mathematical analysis, to name a few. And some of his greatest achievements belonged to the Ramsey theory. This theory was named after the British Mathematician Frank P. Ramsey, a close friend of the philosopher Ludwig Wittgenstein.[10] Although Ramsey died at the age of 26, he contributed greatly to the development of the twentieth century mathematics. Let us see what the Ramsey Theorem deals with, so we can gain a glimpse of the type of thing Erdős was interested in.

Ramsey's Theorem, for which Ramsey is best known, is stated and proved in his paper *On a problem of formal logic*, which appeared in 1928.[11] To understand what it is about, let us start with a pigeonhole principle. This principle states that if there are n number of pigeonholes, and $n+1$ number of pigeons, then one of the pigeonholes must contain at least two pigeons. Right? That's easy.

Now you can go on with increasing the number of these additional pigeons from 1 to infinity. I suggest you make a few intermediate steps

FIGURE 9.4 $n + 1$ number of pigeons in n number of holes, for $n = 9$.

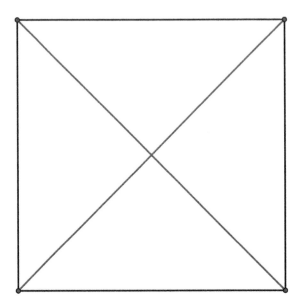

FIGURE 9.5   K4 graph coloured in two different colours. How many different ways can you colour this graph using two colours (only the colours of edges are important)?

first. What happens when you have $n+2$ or $n+3$ numbers of pigeons? And then think about what happens if you have an n, so finite number of pigeonholes, and $n+\infty$ number of pigeons? The pigeonhole principle says that you will end up with at least one of the pigeonholes that will contain an infinite number of pigeons.[12] Ok, we are approaching (but perhaps will never quite reach!) the explanation of Ramsey's Theorem.

Before we proceed, we need to establish how to make a Ramsey number. A Ramsey number is based on a play with coloured graphs. A two-coloured graph is one whose edges are coloured in two different colours. Graphs, by the way, have vertices and edges. So if we look at a simple graph K_4 we can colour that graph in quite a few possible ways. How many? I'll leave that to you to perhaps work out, but here is one such possibility, given in Figure 9.5.

A graph like the one above is said to be complete when every *possible* pair of vertices is connected by an edge. The graph given in Figure 9.5 is labeled K_4, and a complete graph with the n number of vertices will be labeled K_n. Now the Ramsey Theorem's everyday example was formulated by Erdős as follows:

> *Prove that at a party of six people either there are three mutual acquaintances or there are three mutual strangers.*[13]

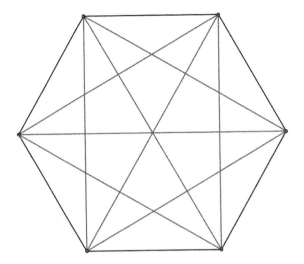

FIGURE 9.6   A possible colouring of K6.

How to represent that with a graph? Let us use what we have learnt so far. Let us first draw a complete graph and specify that the people are vertices (points) and edges (lines that connect them) represent the relationships they have. Let us say that the red edge represents that people know each other or are friends, whilst the blue edge represents that two people that connect them have never met. The theorem really says that in a party of six people there must at least occur one of these cases: that at least three know each other, *or* that at least three do not know each other.

You can extend this and do all the possible colourings if you so wish. Get ready for it though, set some time aside, as you will see that there are many such colourings. Or we can think about it. We can use our pigeonhole principle to make a kind of general conclusion without doing such colouring. Look at any of the edges of the example of K_6 graph as one in Figure 9.6. Each of the vertices has five edges coming out if it – five relationships (as it is a complete graph).

Out of these five edges, as they are coloured either blue or red, we can have either three of them blue or three of them red. This means that there are either three people who are connected with a friendship/acquaintance in this group, or three people who are strangers to each other. It is like going back to that pigeonhole principle – as there are five edges and we say that at least three of them will be one or other colour we can see why that is – we have to get them to somehow fit our graph (like the extra pigeons in their holes).

An interesting thing is that we cannot have a party of less than six people for this to hold. And that is, in fact, what the theorem states. From here, we can then define Ramsey numbers. We say that the **Ramsey number** is written as $R(m,n) = V$ and is a minimum number of vertices such that there would be a combination of at least m or at least n types of sub-graphs or substructures. To translate that into our earlier example, for R (3, 3) = 6, which is the smallest number that gives you the group of people where at least three will either be friends, or at least three will never had met before. This can get a bit complicated, and I'll give you a taste of that via one famous theorem from Ramsey theory. For any given m and n there is a number V such that if V consecutive numbers are coloured with m different colours, then it must contain an arithmetic progression of length n whose elements are all the same colour.[14]

Generally speaking it is said that Ramsey Theory is a system that studies the conditions in which order must appear. Ronald Graham, friend of Erdős, described Ramsey theory as a branch of combinatorics.[15] And to get back to our game of finding the largest number we can think of, that is structured through some kind of system, we can find an example of a problem in Ramsey Theory that gives us a huge number. This number is called Graham's Number. The birth of this number should have featured in the news when it was invented, as it is such a big number, it hurts to think about it. It is defined in a following way: What is the smallest value of n for which every colouring contains at least one sided-coloured complete subgraph on four coplanar vertices in an n-dimensional hypercube? For our example, the coplanar complete subgraph in our illustration in Figure 9.7 is labeled as ABCD.

Instead of planar graphs, Graham introduces n-dimensional hypercubes to obtain a complete graph on vertices, and colours them either blue or red. So Graham asked himself this question: What is the smallest value of n for which every such colouring contains at least one single-coloured complete subgraph on the four coplanar (coplanar means 'in the same plane') vertices?

Are you getting slightly dizzy yet? I hope so as this is indeed a very large number. Even smaller numbers lead to mistakes. I mean not really small, but small compared to Graham's number – like a googol.

A FAMOUS MISTAKE AND THE VALUE OF BIG NUMBERS

The life of googol started when, in 1938, the then nine-year-old boy Milton Sirotta, nephew of an American mathematician Edward Kasner named 10^{100} 'googol'. He proceeded to propose that $10^{10^{100}}$ i.e., 10^{googol} should be called googolplex. But as names and letters go, they always leave some space for

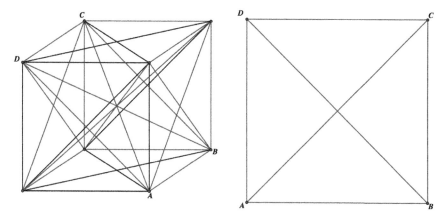

FIGURE 9.7   Graham's number – this is a common example of a three-dimensional cube, coloured with two colours.

spelling mistakes. And so it was that, some fifty years after this event, the inventors of the now most popular search engine in the world wanted to pay homage to this big number but misspelt the name of this number and called their engine Google instead. They missed an 'o'! Easy to do when you have so many of them around one way or another. This spelling mistake, although a little embarrassing, did not have a negative influence on the Google inventors' fortune: Their bank accounts apparently feature many zeros, so many in fact that one more or less would not make too much difference to them.

So if you are tired by now, lying and contemplating the number of grains of sand on the beach of your choice, or how many n-dimensional hypercubes you can construct with two colours in your mind before you pass out, think of this: There is an ever expanding universe of numbers and the definitions of that which is considered to be large seems to be always changing. Tomorrow, both our real, and our universe of numbers, will become just a little bigger. And with friends, as Erdős showed us, all of that will matter even more.

NOTES

1 See Goffman (1969).
2 See the Oakland University Erdős number project page: oakland.edu/enp/readme/ (accessed 1 February 2018). You can also investigate futher at how this collaboration distance is assessed, and the criteria by which the number is given to an Erdős collaborator. MathSciNet (2018) Collaboration Distance, American Mathe-matical Society, https://mathscinet.ams.org/mathscinet/ freeTools. html?version=2 (accessed 19 February 2018).

3 This quote is often attributed to Erdős himself, but Erdős himself attributed it to Alfréd Rényi (1921–1970), another Hungarian mathematician and Erdős's friend and collaborator – they wrote 32 papers together.

4 In 2000, the Clay Mathematics Institute offered a $ 1 million for the first correct solution to the seven Millennium Prize Problems. The prize was awarded some years later to Grigori Perelman, a Russian mathematician living in St Petersburg. He refused to accept it, stating that the greatest prize is in the solution rather than a payment.

5 Ronald Graham is a professor at the University of San Diego in the US. He was one of Erdős's collaborator and a close friend.

6 The Collatz sequence is generated through Collatz conjecture as we showed. It was named by its inventor Lothar Collatz (1910-1990), who first introduced the idea of this sequence in 1937. Erdős is reported to have said that "mathematics may not be ready for such problems". Yet.

7 The first quote is from Pindar (c. 522–443BC) who wrote *Olympic Odes*, and appears in *Ode* II (Pindar, 1997). The second quote is from Aristophanes (c. 446–386BC), appearing in his *Acharnians* (Vardi,?).

8 For the Archimedean system of numbers, see Heiberg (1953).

9 Names of big numbers that were given by Nicolas Chuquet (1455–1488) and reported for the first time in Estienne de La Roche's (1470–1530) book *l'Arismetique* in 1520.

10 Frank P. Ramsey (1903–1930) was an English mathematician and logician, who studied at Trinity College, Cambridge. He was a close friend of Ludwig Wittgenstein (1889–1951), and apart from founding the theory now named after him, was instrumental in translating Wittgenstein's major work, *Tractatus Logico-Philosophicus* into English language.

11 See Ramsey (1928) and also Graham (1983) for further reading.

12 This is also known as Dirichlet pigeonhole principle, after the mathematician Peter Gustav Lejeune Dirichlet (1805–1859) described it in 1834, under the name of shelf-principle or drawer principle (schubfachprinzip). Because of this original name, sometimes the principle is also called Dirichlet's box or drawer principle.

13 This problem became a part of the mathematical folklore, a usual puzzle that is discussed and worked on, since it appeared in the Putnam Competition in 1953 and was posted as a problem in the American Mathematical Monthly in 1958 (Problem E1321). But Erdős, while he worked on it, came up with what became known as a Friendship Theorem, which is not the same as the Friendship graph – that is yet another thing.

14 I would recommend reading more on this theorem, which is called Van der Waerden's theorem, after the Dutch mathematician with the same name (1903–1996).

15 See more from Graham & Butler (2015).

CHAPTER 10

October

THE ROUNDED LIFE AND MUSIC OF VIBRATING STRINGS

If you thought that mathematics is a cold and (only) a rational discipline, you will be sorely disappointed by the story of a well-rounded life of a mathematician who grew from less than humble beginnings to be one of the most celebrated intellectuals of his time. Real, imaginary, and invented, mathematicians come in all guises. Who would have thought that, when a baby boy was left on the steps of the church St Jean le Rond in Paris at the beginning of the eighteenth century, it would grow into one of the most innovative and prolific mathematical authors of his time. During his life, his physical range of influence was small, from the church with the same name as his own, he only moved slightly, mainly retaining the limits of his existence to Paris. But his opus and the circle of his interests were vast. With the introduction of his concept that would change mathematics forever, we draw out a song of d'Alembert's life (it has something to do with the Les Miserables too!). Jean-Baptiste le Rond d'Alembert, French mathematician and scientist, one of the editors of Encyclopédie, died on 29th October 1783.

WHEN I STARTED THIS BOOK, the Notre Dame cathedral in the center of Paris still stood proudly on its little island in the middle of Seine. By the time this chapter of the book was written, the cathedral was badly damaged in the fire and its roof collapsed. The famous novel by the French writer Victor Hugo, *Notre-Dame de Paris, 1482* (published in

FIGURE 10.1   Portrait of Jean-Baptiste le Rond d'Alembert (1717–1783). Colour mezzotint by P. M. Alix after M. de la Tour. Credit: Wellcome Collection. CC BY.

1831) therefore resurfaced in the popular press and media, mentioned as a reminder of this great cathedral and the importance it has had in the life of the French nation. However, there was an event described in this novel which has been overlooked, although not for us, that there was, adjoining the cathedral since the twelfth century, originally another church called St Jean le Rond, where unwanted babies were often abandoned. Although this church was demolished in 1748, it was standing in 1717, the year when, on a cold autumn day on seventeenth of November, a baby boy was bundled into some warm clothes and left there.

In my youth, I came across some images from the encyclopedia to which d'Alembert contributed. I was taken by the variety of human activity that this project attempted to illustrate, describe, and analyze. I am of course talking about the *Encyclopédie ou dictionnaire raisonné des sciences, des arts et des métiers* (*Encyclopedia, or a Systematic Dictionary of the Sciences, Arts, and Crafts*) which was a huge undertaking by Denis Diderot and was co-edited by d'Alembert.[1] The project aimed to show the principles of science and the mechanical arts to a wide public. This was devoted to the same idea as Ephraim Chambers' *Cyclopaedia*, but whilst the Chambers' encyclopedia was written by one man, himself, the *Encyclopédie* was written by many, the first such project written in the world by numerous authors who contributed with articles they were experts in. In here is to be found the bulk of d'Alembert contributions: He wrote almost all of the mathematical articles for the 28 volumes. Through these we can see how important he considered mathematics to be, not only as an intellectual discipline, but its social role too. He for example, divided mathematics into pure and mixed, and then included military architecture and tactics in pure geometry.

Between them, d'Alembert and Diderot wrote 71,818 articles for the *Encyclopaedia* whilst d'Alembert still managed to write original mathematics at the same time. He wrote a book which was a first to give a unified view of mechanics – the science that is concerned by the behavior of physical bodies when they are subjected to physical forces. During the last fifty years of his life, in various correspondence and in form of notes, he also drafted about 5000 pages of mathematics, on various topics such as calculus, mechanics, astronomy, optics, and probability theory.

THE FOURTH DIMENSION

Apart from writing thousands of mathematical articles and pages for *Encyclopédie*, d'Alembert was interested in space and time. We name various things after him, for example, there is such a thing as d'Alembert's principle in mechanics and d'Alembert's paradox in hydrodynamics. And there is d'Alembert equation or d'Alembert formula, which he gave for one dimension space. Here, like in his description of the role of time in the fourth dimension, d'Alembert gives time as an independent variable.

His description of space as a fourth dimension was as follows:

I said earlier that it is impossible to conceive of more than three dimensions. A clever acquaintance of mine believes that one might nevertheless consider timespan as a fourth dimension, and that the product of time with volume would in a certain manner be a product of four dimensions; this idea may be contested, but it has, it would seem to me, some merit, if only because of its novelty.[2]

That clever friend of his may be Lagrange, another great French mathematician, but when this little note was written in 1754, Lagrange was only 19, and there is no evidence that the two knew each other at that time. So it may be that the clever acquaintance d'Alembert was referring to was himself: Was he too embarrassed to suggest something as weird as the fourth dimension perhaps? But in later years, in fact about 43 years later, Lagrange did write about time as the fourth dimension: He wrote about the three dimensions of space and used a letter t to denote time, saying that this is already a principle in mechanics, so we can use it elsewhere. And pronto – the fourth dimension made incursion into our three-dimensional space.[3]

THOUGHT EXPERIMENTS

From here on, the fourth dimension got a life of its own. There were some thought experiments that made things possible for people to imagine this additional dimension. There was an interesting question posed by Möbius: If you had a crystal of a structure which resembles a left-handed staircase, what transformation would you need to do to get the right-handed staircase crystal?[4] Möbius said that this can be done if you pass the object through a fourth dimension to gain its three-dimensional reflection of a kind. Later on in the ninetenth century, Zöllner, a German astrophysicist, repeated this thought experiment but simplified it to include just a circle and a point outside: If you wish to get this point 'into' the circle without cutting the circumference, you will need to get the point to leave a plane in which it is lying in along with the circle, and place it into the circle. In other words, you must 'lift' this point out of its plane and outside of its two dimensions and into the third dimension. For some spatial operations to happen, objects must exit their dimensions, which is what Zöllner concluded.[5]

Zöllner was interested in a number of things and for posterity he is probably best known for obtaining the first measurement of the Sun's apparent

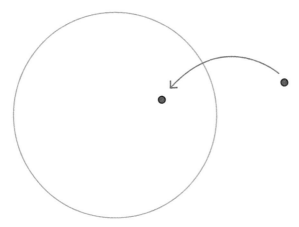

FIGURE 10.2   Zöllner circle.

FIGURE 10.3   Illustration of Zöllner illusion.

size and his work in astrophysics. A lunar crater is even named after him but to the wider public he is certainly best known through his illusion – and if you ever saw it you will recognize it, although you may not know that it was designed by him.

There is something more to Zöllner's thought experiments though. Through his travels to the UK in 1875, he became friends with William Crookes, a Fellow of the Royal Society who was also deeply involved with spiritualism.[6] This movement, which was gathering momentum in England at the time, was a proto-scientific movement that offered an insight into the world which we can only glimpse at and which was inhabited by the souls of the dead. The concept worked something like this: There was a world which is of a greater dimension, the fourth dimension, and all souls inhabit it. Once people have died, their souls have gone from the world of the third

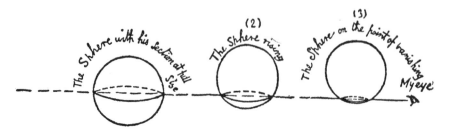

FIGURE 10.4   The diminishing sphere, an object from a third dimension, passing through Flatland, the world of two dimensions.

dimension but they still existed, and could interact with us if we could grasp them in their new habitat of the fourth dimension.

To describe this in mathematical terms, it is perhaps easier to make an analogy of the differences between the worlds of two and three (rather than three and four) dimensions. This is because we can easily imagine a world of two dimensions. A sort of flat-world, a Flatland. Flatland is a land that exists in only two dimensions. It is

> like a vast sheet of paper on which straight lines, triangles, squares, pentagons, hexagons, and other figures, instead of remaining fixed in their places, move freely about, on or in the surface, but without the power of rising above or sinking below it, very much like shadows – only hard and with luminous edges – and you will then have a pretty correct notion of my country and countrymen.[7]

And how would the being from the higher, three-dimensional world appear to the beings in Flatland? Well they would appear as a cross section of themselves. Good thing we have an illustration to show how a three-dimensional sphere looks to the beings of Flatland (Figure 10.4).

There is more of this to discuss, but let us for a moment leave this flat world and look at what that first instance of time as a fourth dimension did to the mathematics of the fourth and higher dimensions.

SOME MORE DIFFICULT MATHS

Don't say I didn't warn you – the subtitle should be a clue about what comes next. You see, if you start using terms such as 'fourth dimension' and then even those that follow you, as Langrange followed d'Alembert, assign a mathematical meaning to it, then you can expect that something will happen further to develop this train of thought. And so it did. We saw some thought experiments about that, but that still didn't take the concept

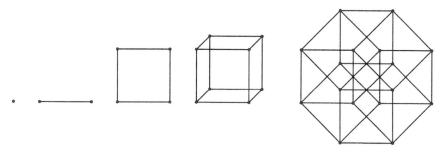

FIGURE 10.5   Starting from zero dimension, represented by a point, we generate a one-dimensional object, the line segment. Further, by moving the line segment perpendicularly to itself, we generate a square, a two-dimensional object. By moving the square perpendicularly to itself, we generate a cube. The cube, a three-dimensional object, moves perpendicularly to itself to generate the four-dimensional hypercube, or tesseract. Image: author.

to its mathematical formulation. At this point it will be useful if we can describe the dimensions from zero to three, which are the ones we can easily imagine.

Starting from zero, we have a point: A point is that which has no part (Euclid, *Elements* I:1). The first dimension can then be imagined as that point moving in a continuous straight fashion: It creates a line, containing an infinite number of points. The line is the first dimension and it only contains points. If we move a straight line in a similarly uniform fashion, perpendicular to itself, we will generate a plane. The plane contains an infinite number of points, lines, and two-dimensional figures. It is a second dimension. Moving a plane perpendicular to itself, we will create a three-fold, or a three-dimensional space. The space contains an infinite number of points, lines, planes, two- and three-dimensional figures (Figure 10.5).

And a late nineteenth century mathematician Stringham suggested that this process can continue into an *n*-dimensional space. How would they look? Well something similar to Figure 10.6.[8]

Studying the fourth dimension became something of a vogue in the nineteenth and the early twentieth centuries. Some very serious mathematics came out of it, for example, one of the most influential mathematicians of the nineteenth century Bernhard Riemann[9] introduced the *n*-dimensional mathematics into mainstream mathematics. There were also many others on the fringes of mainstream academic life and of mathematics, both mathematicians and amateurs alike, who worked on this topic. They tried

FIGURE 10.6 The images above represent 'respectively summits, one in each figures, of the 4-fold pentahedroid, oktahedroid, and hexadekahedroid, with the three-fold boundaries of the summit spread out symmetrically in three-dimensional space' (Stringham, 1880; p. 6).

to imagine what the fourth (and other) dimensions would look like, and how they could be used in all kinds of ways.[10]

An interesting thing about that in particular was that this fourth dimension shook up the foundations of the three-dimensional Victorian society, and included some people who would not otherwise have been given an opportunity to contribute to the development of mathematics due to the social hierarchy of their time. This was in a very subtle and fine way illustrated by Flatland, which aimed at lifting people from the flat world in which men and women have different standing even in their own homes. Aim to lift yourself higher and into the world of higher dimensions, so you can really see what is going on in your world, seems to have been the message of Flatland. To be able to do this, all people needed was a little bit of imagination and some mathematics (Figure 10.7).

A FULL CIRCLE BACK TO LE ROND AND HIS WAY OF FINDING HAPPINESS

Despite his criticisms of other scientists and mathematicians, and reported fallings out with the academicians in both Paris and Berlin, every portrait

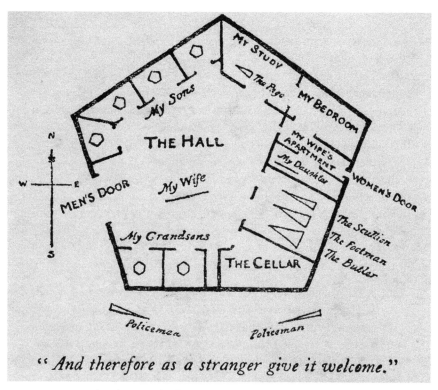

" And therefore as a stranger give it welcome."

FIGURE 10.7   The design of a house in Flatland, with separate doors for men and women. Women, you will notice, don't have angles; they can only be lines. Image: Abbott, 1884; front page illustration.

of Jean le Rond d'Alembert portrays an incredibly happy and cheerful man, watching us knowingly from the distance of history. I searched to see whether it was the glimpse into the possible other dimensions, the mechanics of how the forces of life work, or the fruitfulness of his own life that brought that knowing smile to his face. His life's work and the circle of his interests and contributions to the intellectual and the history of mathematics were considerable. This evidently brought him a huge personal enjoyment, but was there anything else? I continued to search until I found that, perhaps with mathematical precision, d'Alembert studied also the principles of life and found that the only safe harbor he could find would be to keep the journey constant. And happiness? That he said can only be found if a person is able to see the world just as it is, learn how it works, and always maintain the balance between his own interests with fulfilling their duty to humanity.[11]

NOTES

1 The *Encyclopédie* was published between 1751 and 1772 and d'Alembert was its co-editor until 1759.

2 D'Alembert wrote about this in his article on "Dimension" in *Encyclopédie ou Dictionnaire Raisonné des Sciences, des Arts, et Métiers*, Paris, vol. 4, p. 1010.

3 Joseph-Louis Lagrange (1736–1813) was a French-Italian mathematician and wrote this in his book *Théorie des fonctions analytiques*.

4 This question was posed in 1827 – see Lawrence (2015).

5 For further details see Zöllner (1878) and Möbius (1827).

6 William Crookes (1832–1919) went far in his occult activities. He became a president of the Society for Psychical Research in 1890s, joined the Theosophical Society and the Ghost Club and was president of the latter between 1907–1912.

7 Abbott (1884, p. 29).

8 Stringham explained this in the following way:

> *A pencil of lines, diverging from a common vertex in n-dimensional space, forms the edges of an n-fold (short for n-dimensional) angle. There must be at least n of them, for otherwise they would lie in a space of less than n dimensions. If there be just n of them, combined two and two they form 2-fold face boundaries; three and three, they form 3-fold trihedral boundaries, and so on. So that the simplest n-fold angle is bounded by n edges, faces, 3-folds, in fact, by k-folds. Let such an angle be called elementary elementary.*

The fact this doesn't actually look elementary to all, is beyond the point – the angle will be called as such, said Strigham (1880; p.1).

9 Bernhard Reimann (1826-1866) was a German mathematician whose hypothesis, known of course as Reimann hypothesis, is one of the best known papers in analytic number theory.

10 See a paper I wrote on this story in Lawrence (2015) where there are further details in particular on Alicia Boole Stott.

11 Well in fact he said: *Mais ce qui appartient essentiellement et uniquement à la raison, et ce qui en consequence est uniforme chez tous les peuples, ce sont les devoirs dont nous sommes tenus envers nos semblables. La connoissance de ces devoirs est ce qu'on appelle Morale…*

> *Tous ces principes aboutissent à un point commun, sur lequel il est difficile de se faire illusion à soi-même; ils tendent à nous procurer le plus sûr moyen d'être heureux, en nous montrant la liaison intime de notre veritable intérêt avec l'accomplissement des nos devoirs. Jean le Rond d'Alembert*, Essai sur les éléments de philosophie, *sec. vii, pp. 179–80 (1759).*

CHAPTER 11

November

WE MAY BE FRIENDS, BUT I STILL HAVE TO CHECK
WHETHER WHAT YOU ARE SAYING IS TRUE

I may have told you many stories about mathematics and mathematicians thus far, but don't take my word for it! You will need to go and investigate yourself. This is a common practice now, but in the seventeenth century it was a great novelty to have such a place where things are done just in this way: You present a paper and it is then examined thoroughly, just in case you got something wrong. The place I am talking about is of course the Royal Society in London. It was founded in November of 1660, followed by a lecture at Gresham College by Cristopher Wren who we met in March. Wren and his friends, Robert Boyle and John Wilkins among others, took their motto to be 'Nullius in verba', literally meaning 'take nobody's word for it'. This is useful advice to bear in mind when one thinks about mathematics and about truth. It gives you an inkling to what mathematics is and the reason why mathematical proofs are examined, re-examined, taught, and learnt. It is not enough, in mathematics, for one to take things for granted, to know mathematics is to know how the proof works and discover it through the 'doing of mathematics'. The Royal Society, the oldest national scientific institution in the world, was founded on 28th November 1660.

WHEN WE LIVE WELL for a long time in peace and relative prosperity, we tend to forget how bad things can be when there is a war going

A MEETING OF THE ROYAL SOCIETY IN CRANE COURT (*see p.* 106).

FIGURE 11.1   Royal Society, Crane Court, off Fleet Street, London: a meeting in progress, with Isaac Newton in the chair. Wood engraving by J. Quartley after [J.M.L.R.], 1883. Credit: Wellcome Collection. CC BY.

on. And perhaps a civil war is worse than any as it is difficult to suddenly find that the friends and neighbors you lived with until yesterday overnight become your deadly enemies. The Royal Society was born out of such troubled times. Although the English Civil War officially lasted between 1642 and 1651, the consequences of this war were far reaching and the peace was by no means to be taken for granted. The Protectorate, in effect the time when the UK was governed as a republic (England and Wales, Ireland, Scotland, and the overseas territories) was under the leadership of Oliver Cromwell until 1658 and his son for another year. The monarchy was restored in the same year in which the Royal Society was founded and King Charles II bestowed upon it the royal charter.

The foundations for Royal Society were already set before this though. In the late 1640s, Robert Boyle wrote to some of his friends about men being organized across the country in some kind of 'invisible college' working on various scientific experiments.[1] The precursor of the Royal Society, this 'invisible college', had in many instances overlapped with the Gresham College, a fully operational institution, which was founded half a century earlier, in 1597.[2]

A group of twelve men from Gresham College met on 28th November 1660 in the rooms of the college in central London, after the lecture by Christopher Wren. In this meeting the new organization was established, and this date, and the meeting, are regarded as the beginning of the Royal Society. The founding members of the Society came from different political groups and were of different persuasions – this seems to had been intentional. At first they called the new organization the Philosophical Society, but later, as they were granted three royal charters, it became known as The Royal Society. If you now say that someone was an original member of this Society, that would include all of its members up to the time when the second royal charter was granted in 1663.

What can we learn from this diverse group? The rebuilding of the country after the civil war they envisaged (and not only their Society) could not begin until the differences of opinion were accepted as something that gives rise to progress, and in a most productive way. They seem to have agreed to disagree in order to find truths science and mathematics can lead them to.

NEW WORLDS ON THE HORIZON

When Amerigo Vespucci showed that the mid- and south-Americas were not the eastern periphery of Asia, which became known after Columbus' discovery of the New World, the new continents were named after him. He never visited America. Another geographer, George Everest some centuries later, who surveyed the majority of India during his working life, became the Surveyor General of India towards the end of his career. He had the greatest mountain in the world named after him. Mount Everest, named in his honor just a year before he died, was never visited by George Everest.[3]

A mathematician can, likewise, discover or grasp the existence of new mathematical worlds too but not necessarily visit or even experience them. When this happens they can perhaps imagine a land that is somewhere out there, over the visible horizon, of what is not yet possible to do with the existing mathematical methods. Such was the case with George Boole, the Fellow of the Royal Society, and the inventor of Boolean Algebra, which lies at the basis of all programming languages and gives the fundamental principles to make possible all of the digital electronic inventions of the past century.

THE BOOLE FAMILY AND THEIR FRIENDS

What follows is a peculiar story of friendship and family in the midst of a bigger story that relates to the invention of Boolean Algebra and its inventor. In this story one of the central people is Mary Everest, later Boole. Mary was a niece of George Everest, a daughter of his brother Thomas Roupell. Mary's uncle, John Ryall was a professor of Greek at Queen's College, Cork. It was through Cork that George Boole met Mary, they married and had five daughters.[4] I'll have to slow down, as in between their meeting and their daughters growing up, George came up with a wonderful system that would take us beyond that visible horizon of the then existing mathematics, and into a world of computers that we so comfortably inhabit now.

George Boole became a professor of mathematics in Cork, having had no formal mathematical training, which was very odd even then. How did that happen? In most articles and books about him, you can read that he was self-taught, and then became a school teacher. What actually happened was that he was forced to find employment at sixteen, in order to help his family. The options were to become apprenticed to his father as a shoemaker, or to gain other employment; this 'other' employment was a post of a teaching assistant at the Methodist school in Doncaster. While there, he became engrossed in the study of mathematics. There are several reports that it was there, while he was crossing a field near the school in which he worked, that he had a kind of revelation that changed the history of mathematics and technology forever. He reportedly had a kind of an epiphany and at once understood that the human mind has a propensity to unify all experiences and reorganize them as it learns new things. This 'innate sense of unity' led further to structuring his mathematical thought and his new science of logic by which something can either be expressed as 1 or 0. All or nothing.[5]

George changed between a few schools and ended up starting his own in Lincoln at one stage. During this period, he progressed further in his study of mathematics and began to publish articles in the *Cambridge Mathematical Journal*. Boole introduced himself to the first mathematics professor of the University College, London, Augustus de Morgan by a letter in 1842.[6] De Morgan secured a pass to the library of the British Museum for Boole (which became the British Library, but was at the time positioned in the Round Room of the main Museum building), which must have been useful on his occasional visits to the capital. Perhaps not an important fact in itself, it shows that Boole did not himself have access to such sources as could be found in this great library, and which would have been crucial for his research, until he met De Morgan.

When Boole sent De Morgan one of his papers in 1844, the paper's significance was recognized as a work of genius, and the paper was published in the *Transactions of the Royal Society*. Boole received the Society's Royal Medal for this paper and his career rocketed.[7]

Exactly ten years after this paper, George, who had by this time became a trusted friend of the Ryall and consequently Everest family, published *An Investigation into the Laws of Thought, on Which are founded the Mathematical Theories of Logic and Probabilities*. The book he dedicated to John Ryall displayed a dedication 'in testimony of friendship and esteem.' This was a seminal work in mathematics, logic, and later in computer science. It was immediately recognized as being extraordinary but people of course could not at the time predict that it will be an invaluable contribution to the development of digital technology. That recognition only came in the twentieth century when a US master's degree student Claude Shannon showed how Boolean algebra can be used to construct digital circuits. He showed this on an example of automatic telephone exchanges.[8] Since then, everyone knows that there are 10 kinds of people: those who understand binary and those who don't!

AS MATHEMATICAL FAMILIES GO...

George's good friend, John Ryall, was Mary's uncle from her mother's side. That is how George met Mary, when she visited her uncle in 1850. Mary was, by all accounts of those around her, and through her own extensive writing about her life, a very interesting person.[9] The only daughter to her parents, she spent some of her childhood in France when her father wanted to work with Hahnemann on the new pseudoscientific system of medicine – homeopathy. Mary's father had already written a book about homeopathy and Hahnemann was known as 'the father' of the discipline.[10] It was here, in France, in a castle near Paris, that Mary came across mathematics and fell in love with algebra. She reported how she identified working on algebra as a kind of sacred experience, a kind of a door into the world where she could (quite literally in her case) converse with God.

Mary and George married upon the death of her father and were by all accounts a very happy couple. They had five children, all daughters, but after six months of the birth of their last child, George Boole died. Everyone (most probably herself too), blamed Mary for it, as she applied one of the homeopathic principles of her father that states that 'like cures like' and applied cold wet sheets to her husband's body as he developed a cold chill after being drenched in a downpour. George as a consequence caught pneumonia and died. In her later life Mary become an active spiritualist. Like many spiritualists, she too had an irresistible urge to get in touch with the dead, and in her case this may have been linked to a hope that her husband would grant her forgiveness.[11] Mary it seems, as her youngest daughter Ethel testified in a letter written many years later to her nephew Geoffrey (Geoffrey Ingram Taylor who also became a member of the Royal Society), had a propensity for 'cranks.'[12]

Mary was inclined, as was her father, to being interested in exploring new methods to study various phenomena related to the relationships between science and the spiritual world in whichever way they would have been defined. One should not forget that this was the time too of the great rise of so called 'occult' sciences. The many new scientific discoveries and inventions, such as electricity and magnetism, certainly played their role in igniting a new thirst for the further investigations of hidden structures and forces, and sought to define a science of the spirit.[13] Although the majority of people and methods they used lacked the precision to lead to new inventions in the fields of real science and mathematics, they in a very around-about way contributed to creating a fertile ground for creative and new thinking. Through her interest in spiritualism, Mary Boole became a friend of James Hinton after the death of her husband. Hinton, although a famous and public polygamist, was a surgeon and a prolific author on diverse subjects which in some ways demonstrates the origins of his various alternative views on medicine. He was also, most importantly for our story, the father of Charles Howard Hinton.[14]

Charles became a regular visitor to Mary and her five daughters. He was interested in higher dimensions and published a book on how to visualize the fourth dimension. He coined the word tesseract and taught young Alicia Boole, the Boole's middle daughter, the art of seeing the objects, in particular the regular polyhedra of the four dimensions. He apparently taught her to see how such objects would exist in the fourth dimension as well as what would happen when they intersected with the third dimension. Remember Figure 10.4? When a three-dimensional sphere intersects with the two-dimensional flat surface, it leaves a 'trace' which is a circle. Similarly, the intersection of the fourth-dimensional object would leave 'traces' as they pass through the third dimension. Hinton's method of visualizing both the four-dimensional objects and their 'traces' in three-dimensional space was new. He suggested to the learners of mathematics of the fourth dimension to give up on conceiving the fourth dimension itself. He is quite clear that this is not possible. What is possible however, is to consider the four-dimensional objects passing through our three-dimensional world, and from this passage and in fact from many such 'projections', to reconstruct the original four-dimensional object.

His method was, it would seem, very successful, although perhaps only with his student Alicia. Hinton married Alicia's eldest sister Mary and they lived in the US after an exploration of polygamy on his part. Hinton died suddenly during a lecture. The anecdotal history tells that previously in his talk he suggested he will visit the fourth dimension and duly fell dead at the scene.

But Alicia lived a long and happy and very productive life. As a child she had already had experience of mathematics through her father. Although she was only six when he died, one could imagine that they had shared mathematical interactions, in addition to her conversations with her mother. Mary published a number of books on mathematics, in particular on how to teach the discipline and inspire young minds. Later in life, Alicia, through her discussions on the nature of space with her husband, Walter Stott, learnt of the work of Dutch mathematician Pieter Hendrik Schoute.[15] Schoute worked on central sections of the regular polytopes. Alicia never stopped experimenting with the fourth dimension since she was taught how to visualize regular polytopes by Hinton, and was making models of three-dimensional projections of four-dimensional polytopes. She sent pictures of these models to Schoute in 1895, which amazed him. This was the beginning of a very long and productive cooperation: They worked together for twenty years, during which she published papers on her own and with Schoute. Alicia was given the honorary doctorate by the Schoute's University, University of Groningen, upon his death.[16] From 1930 she worked with Donald Coxeter, an English mathematician. Donald Coxeter (1907–2003) was a mathematician specializing in geometry. He was born in London and gained his PhD in Cambridge, but lived for most of his life (from the age of 29) in Canada, where he was a professor of mathematics at the University of Toronto.

There are families, as it were, in which the learning of mathematics runs. The Booles are certainly something special. From father George and mother Mary, Alicia became a mathematician who, again without a day of official mathematical training, was awarded an honorary doctorate in mathematics although she never gained an official position.

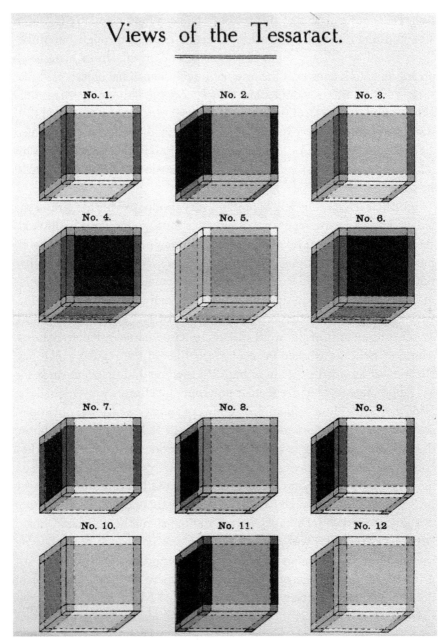

FIGURE 11.2 Hinton's illustration of the tesseract and how it exists in this higher, fourth dimension. Hinton, 1888, frontispiece.

THE LEGACY FOR THE FUTURE

The Royal Society currently has around 1600 fellows and foreign members. Over its history, it has had more than 8000 fellows, including the most famous mathematicians and scientists: Newton, Darwin, Faraday, Einstein, Turing, Hawking were all members. Some of them are very famous and all have moved scientific and mathematical understanding beyond the existing boundaries and into the new fields of knowledge and understanding of the world. But my attention has been gripped by George Boole for several reasons. His humble beginnings, his incredibly modest life, and his work that become interesting again about a hundred years after its invention. His family, whilst unorthodox in more ways than one, was an incredibly interesting family who gave the world mathematicians, politicians, and writers. George's daughter Ethel for example, wrote the novel *Gadfly*, which was a bestseller in the Soviet Union and China where it sold more than five million copies. George and Mary also gave to the world two important mathematicians, his daughter Alicia, and his grandson, another Royal Society Fellow, Geoffrey Ingram Taylor.[17]

Beside how interesting the story of his and his family's life has been, Boole was particularly interesting because of his new mathematics. He delved where few dared to go: He looked at that most fascinating subject of finding the underlying structures of human thought. And to understand and systematize his findings he invented a new mathematical technique, the Boolean Algebra. Based on his explorations, Boole succeeded in outlining the methods and structures that enabled, almost a century after his work, for a work to begin on developing machines based on the principles he established. Machines are in turn, another couple of centuries after Boole, beginning to learn themselves.

NOTES

1 Robert Boyle is one of my favorite scientists from this time of the founding of Royal Society. Not only was his science inspiring, but he also wrote *The Martyrdom of Theodora and Didymus*, which was later taken as the basis for the libretto of *Theodora*, an opera by Handel.

2 Sir Thomas Gresham the Elder (1519–1579) was an Elizabethan merchant and financier. He founded the Royal Exchange in the City of London. He bequeathed majority of his estate to his wife with the stipulation that after her death their house in Bishopsgate Street should be used to establish a college. Thus the Gresham College was founded in 1597.

3 Some say that this was a kind of compromise as several local names were put forward, and the British administrator of India at the time, a successor of Everest, Andrew Scott Waugh (1810–1878) put Everest's name forward.

4 Some dates for orientation are needed here: George Everest (1790–1866), Thomas Roupell Everest (1800–1855), Mary Everest, later Boole (1832–1916), George Boole (1815–1864). The five daughters of Mary and George were Mary Ellen (1856–1908) who married Charles Hinton (1853–1907) and of whom more later; Margaret (1858–1935) whose son Geoffrey Ingram Taylor (born 1975) is also mentioned later in this chapter; Alicia (1860–1940) who worked on visualizing the fourth dimension and is also mentioned later in this chapter; Lucy Everest (1862–1904), and Ethel Lilian (1864–1960) who wrote *The Gadfly*, which became a bestseller in the Soviet Union and the People's Republic of China.

5 Now this was reported after Boole's death by his wife Mary, so we can't be 100% certain that it is exactly how things happened; the belief is that this has happened. See for example Cohen (2007) chapter 3 and Kennedy (2016) also chapter 3.

6 Augustus De Morgan (1806-1871), was the first professor of mathematics at the University College London.

7 This paper's title was *On a general method of analysis* and consisted of his method to apply algebraic methods to the solution of differential equations. Boole got the medal in 1844, and was appointed the Chair of Mathematics at Queen's College Cork in 1849.

8 Claude Shannon (1916–2001) was an American mathematician and cryptographer, and has often been credited with the invention of Information Theory. His thesis came under the title *A Symbolic Analysis of Relay and Switching Circuits*, was written in 1937, and can be found on the MIT website at https://dspace.mit.edu/handle/1721.1/11173 (accessed 1st July 2019).

9 Quite a lot of her writings can be freely accessed as digital copies at the Internet archive under her name.

10 Mary's father Thomas wrote a book *A Popular View of Homeopathy* in 1834. He came across a Samuel Hahnemann (1755–1843) who experimented with homeopathy and is often credited with the invention of homeopathy. Samuel gained the following in Thomas who then took his family for a three year stay to France, and they lived in a Chateau de l'Abbaye at Poissy near Paris while he worked with Hahnemann.

11 About the links between science and the spiritualism in Victorian Britain, see Noakes (2004).

12 Quote from the one of her letters to her nephew Geoffrey Taylor, who was a mathematician in Cambridge; the letter can be found in the Trinity College, Cambridge, *Papers of Sir Geoffrey Taylor*, A131(14). This was mentioned in Kennedy (2016), p.65.

13 For the history of the Psychical research, see a mathematical historian's account in Grattan-Guinness (1982).

14 Charles Hinton (1853–1907) was a British mathematician who became interested in the fourth dimension, and was the one who coined the word 'tesseract' for the four-dimensional analogue of cube. He was married to Mary Ellen, Mary and George Boole's eldest daughter. Hinton studied at Balliol College Oxford, and in later life lived and worked as an academic and an examiner of patents in the US.

15 Walter Stott (dates unknown) was an actuary. Pieter Hendrik Schoute (1846–1923) was a Dutch mathematician who worked at the university of Groningen.

16 Although she accepted, she did not travel to receive the award, which was then given to her 'in absentia'.

17 Taylor became a major figure in developing the theory of fluid dynamics, among other things.

CHAPTER 12

December

WHAT CAN A MATHEMATICIAN GIVE TO THEIR FRIEND AS A NEW YEAR'S PRESENT?

Mathematics rewards the mathematicians and their friends in most unexpected and beautiful ways. Johannes Kepler, mathematician and scientist, was born on 27th December 1571. Kepler is the last mathematician we visit on our journey. On one cold December night, around his fortieth birthday, Kepler was walking around the snow covered streets of Prague. His friend and patron, Baron Johannes Matthaeus Wacker von Wackenfels was due a present for New Year's Day. Kepler's own birthday may have been an inspiration – did Kepler receive a present from Wackenfels on that evening and on his way home was wondering how he could reciprocate? What could he give to his friend in return?

IT WOULD BE NICE, if somewhat cold, to follow Kepler on this night stroll through the snow covered Prague towards the very end of 1571, and see him enchanted by the beauty of life and by geometry that was contained in snowflakes. As he touched the snow it melted and the delicate shapes (which can sometimes be seen by a naked eye) would disappear. Kepler asked himself why each snowflake, while different in detail, was of the same regular hexagonal shape? Could there be a snowflake which would be three-, four-, even seven-sided? The thoughts that began to form on that evening's walk Kepler soon put to paper. This first initial idea led him to think not only about snowflakes, but other kinds of matter and how they

FIGURE 12.1   Johann Kepler (1571–1630). Line engraving by F. Mackenzie. Credit: Wellcome Collection. CC BY.

are formed and organised. The most beautiful thing, he mused, was the shape of matter, *not* its content. His friend, Wackenfels,[1] who he called a 'devotee of Nothing' would be the person to value such a study more than any material wealth. He dedicated his little work *On the Snowflake, or the six-sided crystal* (*Strena Seu de Nive Sexangula*) therefore as the "New Year's gift for the devotee of Nothing, the very thing for a mathematician to give, who has Nothing and receives Nothing, since it comes down from heaven and looks like a star"[2]. Perhaps, the pursuit to understand this 'Nothing' may be, after all, much more valuable than any other present we may be tempted to give to a friend. This Nothing was of course, space, with spatial structures in its many manifestations, and possible to understand by those who study mathematics.

THE JOURNEY TO PRAGUE

Let us interrupt this story for a moment to see Kepler journey so far from his youth to this moment. Young Kepler, who was born in the, as was then called, Free Imperial City of Weil de Stadt, in the Stuttgart Region, now Germany, studied theology. He was preparing for a priesthood at the University of Tübingen, a life that was not to be. He must have in some way impressed one of his professors, Michael Maestlin[3] by his understanding of astronomy or mathematics. Kepler did not recognize how talented he was, of which he wrote some years later perhaps unsurprisingly, as his worst grade at the university was in astronomy.[4] His university Senate stated that he had a magnificent mind for astronomy and he was dispatched to become a teacher at a Lutheran provincial school in Gratz, now Austria.

While a teacher there, Kepler studied Plato and stumbled upon the five regular solids featured in the *Timaeus* (c.360 BCE). You will remember from earlier chapters, that these solids, now often called Platonic, were connected with the four classical elements: earth was represented by cube, air by octahedron, water by icosahedron, and fire by tetrahedron. The fifth element, the dodecahedron, was believed to have been used for arranging the constellations of the heavens. This was often also identified as the element denoting a divine spark, the principle of attraction, the force that made all other elements come to life.

Thinking of these five solids, Kepler came up with a question that had previously begged to be asked: Could these solids be nested in one another? What would that look like and how could that be constructed? Perhaps he experimented with physical models? We don't know, yet we have an image that was published in a little treatise he put together, his *Mysterium cosmographicum*, or *Sacred Mystery of the Cosmos*, printed in 1596. This, probably one of the most famous images from the history of science and mathematics, represented his model of the universe. It shows each Platonic solid encased in a sphere, inscribed in a further solid, encased in a sphere, which Kepler identified as the then six known planets from the solar system: Mercury, Venus, Earth, Mars, Jupiter, and Saturn (see Figure 12.2). These are assigned solids in order from octahedron, icosahedron, dodecahedron, tetrahedron, and cube from the center outwards. The spheres containing the solids are placed at intervals corresponding to the sizes of each planet's path as they were then known, assuming that they circled around the Sun. This in itself was revolutionary in those days, as it was only less than a half a century previously, in 1543 to be precise, that the established view of the structure of the solar system was challenged, when Nicolaus Copernicus (1473–1543) a Polish astronomer and mathematician, showed that planets revolved around the sun, rather than the sun, moon, planets, and stars around the fixed Earth.

Of course, later, Kepler found that the orbital paths of planets of the solar system were not circular but elliptical, but his beautiful model, even though it was not perfectly accurate, gave him an impulse for further research. Kepler sent his little book with a friend who was traveling to Italy, with the instruction to give copies to interested persons. One of these copies was given to Galileo who consequently corresponded with Kepler about it.

FIGURE 12.2 Photograph of a model of Kepler's *Mysterium Cosmographicum* main illustration. Credit: author.

Another came into the hands of a great Danish astronomer Tycho Brahe, who left Denmark for Prague to establish an observatory for Rudolph II, the then King of Bohemia.[5]

Kepler was invited to join Tycho, which came at a good time as the movement that became the Counter Reformation gathered pace in German lands, and Kepler felt his stay in Gratz was becoming, frankly, uncomfortable. Initially this was a war between Protestants and Catholics, but it ended up getting more and more complex, and after thirty years (a Thirty Years' War as it is now called) and more than eight million deaths, it ended. Witch hunts were prevelant during the war: When all else failed to accuse the innocent and kill them in the most horrific ways, people would come up with an accusation that the so-and-so was a witch.

One can then forgive Kepler for leaving Gratz and going to Prague. This took place at the beginning of 1600, when he first met Brahe and his assistants at the site where the new observatory was being constructed, near Prague in Benátky nad Jizerou. Kepler went back to Gratz to gather his family with the view of moving away, but while there in August 1600, he was exiled from the city for refusing to convert to Catholicism. If he had, until then, any doubts whether to abandon Gratz forever, they were now quashed. Kepler came back to Prague where he would become the imperial astronomer and mathematician after the untimely death of Brahe in October 1601. The following eleven years would be some of the Kepler's most productive. During that period, he would publish his *Astronomiae Pars Optica* (The Optical Part of Astronomy, 1604), *Astronomia nova* (A New Astronomy, 1609), and of course his *Strena Seu de Nive Sexangula* (A New Year's Gift of Hexagonal Snow, 1611). This period came to a close with the abdication of Emperor Rudolph in favor of his brother Matthias, and Kepler's family's poor health and death of his six-year old son Friedrich. Kepler sought to move with his family elsewhere, but in the meantime his wife also died as did Emperor Rudolph. At this point Kepler was confirmed as the imperial mathematician by Matthias, and upon request was granted dispensation to move to Linz in 1612.

THE WITCHES AND DREAMS

Kepler's laws of planetary motions and through these his modeling of the Solar System were based on his mathematics, but also on his deep religiosity. Behind this was his deep insight into how the world works. This is beautifully illustrated by the imagery in his work. His lesser known book, a little novel about a dream that gives the dreamer knowledge of how the Solar System works, is both strange and mystifying. In the *Dream* (originally published under the title *Somnium*) the main character, an Icelandic boy takes flight around the Solar System guided by his mother. The boy is called Duracotus and he has a warning for his readers: "While my mother lived she would not let me write: She warned of evil men who reviled the hidden arts because their dull minds could not grasp them, who spread lies and made laws that harmed the human race…"[6]

The haunting nature of this science fiction novella makes even a more gripping effect on the contemporary reader as we find that Kepler could not actually publish this book, as his mother was accused of witchcraft in 1617. As I mentioned earlier, during the Thirty Years' War it was not uncommon to accuse people of witchcraft. It was, however, very unusual for them to be defended and released. The coincidence of various commonalities between the reality of Kepler and his mother's life, and Kepler's novella, the *Dream*, was quite astounding. His mother was accused of witchcraft in 1617, eleven years after he had begun writing the *Dream*, in 1608. She was released some years later, in 1621, mainly because of Kepler's own efforts and involvement in the trial. He proved that she could not be a witch. But imagine if the novel was published! I neglected to mention that the boy Duracotus and his mother Fiolxhilde were helped on their journey around the Solar System by a flying demon…

Of course, this all looks terribly suspicious to an unsuspected reader and you may even be tempted to make a conclusion that both Kepler and his mother were really witches and/or wizards! But the story is a little more complicated as life usually is, and my earlier dates have an additional set of information that makes this whole story clearer. Kepler actually wrote in 1608 the outline of the story in which he suggests that an observer standing on the Moon would be able to see clearly the movement of Earth. Over the next decades, one may imagine through a good doze of humor, he added the structure which turned a scientific explanation into a story. As the events of his mother's trial unfolded, Kepler reflected on the turbulent years of the witch trials as well as the meanness of the people he came across during that period.

FRIENDSHIPS THAT LAST CENTURIES

Kepler, like all of us, learnt from those who came before him. He worked on the Platonic solids and looked at those other, semi-regular polyhedra, Archimedean solids, of which we have spoken about in June, in the chapter on Pacioli. In fact, it was Kepler that consolidated this work which included contributions from Archimedes, through to late antiquity mathematicians, and then some centuries later, the Renaissance mathematicians as Pierro della Francesca, Luca Pacioli, and Dürer. It was also Kepler that gave all the images for each of the Archimedean solids and gave them their modern, current names as in Figure 12.3.

However, Kepler did not stop there and he rediscovered two of the four concave regular star polyhedra. The full set of four are now called Kepler-Poinsot solids; they are composed of regular polygons (see Figure 12.4).

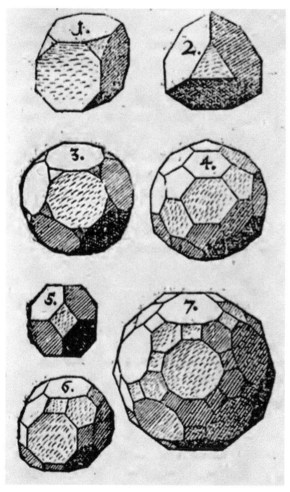

FIGURE 12.3 Archimedean polyhedra, as they appear in Kepler's book, *Harmonices mundi libri V*, Linz, 1619 Book 2, Proposition 28. The names, as in figure 38, are given by Kepler (1571–1630). 1 – truncated cube, 2 – truncated tetrahedron, 3 – truncated dodecahedron, 4 – truncated icosahedron, 5 – truncated octahedron, 6 – truncated cuboctahedron, 7 – truncated icosidodecahedron,

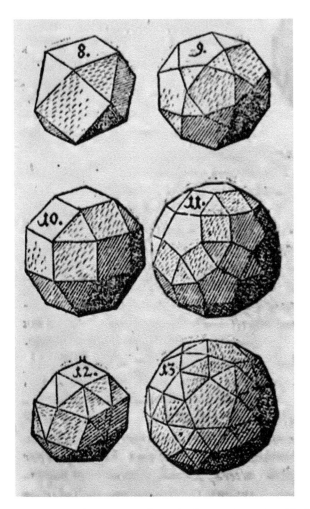

FIGURE 12.3 Continued: 8 – cuboctahedron, 9 – icosidodecahedron, 10 – rhombicuboctahedron, 11 – rhombicosidodecahedron, 12 – snub cube, 13 – snub dodecahedron.

FIGURE 12.4   The small stellated dodecahedron Kepler named as Ss, and great stellated dodecahedron Tt.

LET'S PACK OUR PRESENT NEATLY

There is so much to be written on these and the other regular (Platonic) and semi-regular (Archimedean) solids. I may decide to write about that for some other eventful day in the year. For the moment, let us just conclude that his little paper on snowflakes was not something out of the ordinary – Kepler was already an ultimate expert on these starry and perfect shapes.

Let us now get back to our snowflakes lest they disappear from our imagination and melt into an old memory. Kepler's study on snowflakes, while it began there, turned into a generalization of a theory that Kepler had developed while trying to show why snowflakes are hexagonal: He looked at how spheres pack in the three-dimensional space. This in turn took him to the understanding of how hexagonal shapes are the most efficient ways of packing in a plane. Bees of course have known this all along (Figure 12.5). But Kepler, when a few snowflakes fell on his coat as he

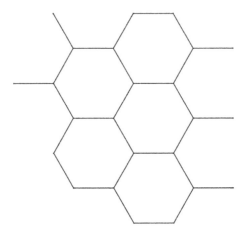

FIGURE 12.5   A honeycomb structure.

was walking around snowy Prague that distant December night in 1611, thinking about what kind of present to get for his friend who didn't need anything, and who most of all liked nothingness, realized that the mathematics of snowflakes themselves must certainly be the best seasonal present for such a friend!

Kepler said in the dedication of his little treatise on snowflakes that he was "well aware how fond you are of Nothing" and that all goods that can be thought of are nowhere near to the beauty of the understanding of nothingness. His gift was instead the discussion on the structures of crystal little stars that fall on us as we pass through a snowy night.

Is this the end of this story? Of course not. We now systematize snowflakes and ice formations and study them in many ways. His study of the packing of equal spheres by closely packing them as if in a crate will be best done if you stagger them in layers, just like greengrocers do, in cubic or hexagonal way. The suggestion that this packaging is most efficient is now called the Kepler conjecture and was proven by Thomas Hales (born 1958) in 1998. Whilst Kepler wrote about the three-dimensional packing, this has been proven to be equally valid for 1, 2, 3, 8, and 24 dimensions. Kepler did mention at the beginning of his treatise the wisdom of bees, and this too has been studied by Hales and is now accepted as a *Honeycomb Conjecture*.[7]

Kepler's little present connects many different areas of mathematics, and many people across the continents and centuries. Like bees, we humans seem to gather together around such ideas to form patterns that collect the dew of our collective intellect. But never forget, that, like every drop of

FIGURE 12.6 Scheuchzer, 1731–1733: *Snowflakes*. Credit: Wellcome Collection. CC BY.

water, this precious dew can evaporate when it heats. Perhaps it is best to keep very cool thoughts and think of snowflakes.

NOTES

1 Johannes Matthaeus Wacker von Wackenfels (1550–1619) was a diplomat and a scholar.
2 Kepler (1975) p. 1.
3 Michael Maestlin Michael Maestlin (1550–1631), German astronomer.
4 Kepler (1937: 3, p. 108). History of science and mathematic has plenty of resources for us to learn more about Kepler. But for the start, recommended reading would be Gingerich (1970–1990).
5 Tycho Brahe (1546–1601) was a Danish nobleman and astronomer. After he left Denmark in 1597, he was invited by the Holy Roman Emperor Rudolph II (1552–1612) to come to Prague. There Tycho became the official imperial astronomer, but died shortly afterwards, in 1601.
6 *Somnium*, or *The Dream*, is a novella written by Kepler over a period of years and published posthumously by his son in 1634. It was written in Latin and has been very much liked as both science fiction and a scientific treatise on lunar astronomy.
7 Honeycomb Conjecture states that a regular hexagonal grid just as the one made by the bees around the world, is the best way to divide a surface into regions of equal area with the least total perimeter. Although the conjecture was referred to by Pappus around 325-340 AD in his Book V, it was only finally proven by Hales (1999, 2001).

Final Remarks

A HAPPY NEW YEAR'S CARD

Here we conclude with a short note, with very good wishes for your continuation of the study of all the beautiful things that mathematics brings and can give us insights into. Perhaps we've seen a greater share of weird interests and beliefs than usual, in particular with mathematicians, but this was not done on purpose or with any ultimate motive. The mathematics we witnessed through these stories, was both new and old; of this, and of the other, imagined or real, immemorial and sometimes future worlds. As these mathematicians have demonstrated to us, this kind of meddling, investigating, and dreaming of mathematics, when creatively combined, can produce some amazing new things that bring progress and benefit all human kind. With that thought I hope I have given you some images with which to paint the mathematics that you will want to learn and make in the future.

Bibliography

Abbot, E.A. (1884). *Flatland*. London: Seeley & Co.

Alexander, A. (2011). *Duel at Dawn*. Cambridge, MA: Harvard University Press.

Anderson, C.A. (2015). "Gödel's 'Proof' for the Existence of God", in *Mathematicians and their Gods*, Lawrence, S. and McCartney, M. (eds.), Oxford: Oxford University Press.

Apollonius, P. (1566). *Conicorum libri quattuor*.

Archibald, R.C. (1950). "The First Translation of Euclid's Elements into English and its Source", *The American Mathematical Monthly*, 57: 7, pp 443–52.

Ball, W.W.R.(1889). *A History of the Study of Mathematics at Cambridge*. Cambridge: Cambridge University Press.

Baltzly, D. (ed.) (2009). *Proclus: Commentary on Plato's Timaeus: Volume 4, Book 3, Part 2, Proclus on the World Soul*. Cambridge: Cambridge University Press.

Billingsley, H. (1570). *The Elements of Geometrie of the most auncient Philosopher Euclide of Megara*. London: John Daye.

Bourbaki, N. (1939). *Élements de mathématique*. New York: Springer Verlag.

Brewster, D. (1855). *Memoirs of the Life, Writings, and Discoveries of Sir Isaac Newton*. Edinburgh: T. Constable and Co.

Bronowski, J. (1956). *Science and Human Values*. New York: Harper & Row.

Burnett, C. (1987). *Adelard of Bath. An English Scientist and Arabist of the Early Twelfth Century*. London: Warburg Institute.

Burnett, C. (1999). *Adelard of Bath: Conversations with his Nephew*. Cambridge: Cambridge University Press.

Calis, R., Clark, F., Flow, C., et al. (2018). "Passing The Book: Cultures of Reading in the Winthrop Family, 1580–1730", *Past & Present*, 241:1, pp. 69–141.

Cassirer, E. (1943). "Newton and Leibniz", *The Philosophical Review*, 52: 4, pp. 366–91.

Chipp, H.B. (1968). *Theories of Modern Art: A Source Book by Artists and Critics*, Berkeley: University of California Press, pp. 222–23.

Clagett, M. (2001). *Greek Science in Antiquity*. New York: Dover Publications.

Clucas, N. (2008). *John Dee's Natural Philosophy: Between Science and Religion*. London: Routledge.

Cohen, D. J. (2007). *Equations from God: Pure Mathematics and Victorian Faith*. Baltimore: John Hopkins University Press.

Colson, J. (1801). *Analytical Institutions in Four Books: Originally Written in Italian, by Donna Maria Gaetana Agnesi*. London: Taylor and Wilks.

Crépel, P. (2005). "Jean Le Rond d'Alembert", traité de dynamique (1743, 1758), in I. Grattan-Guinness (ed.), *Landmark Writings in Western Mathematics 1650–1940*. Amsterdam: Elsevier, pp. 159–67.

D'Alembert, J. le R. (1751). "Dimension," in *Encyclopédie ou Dictionnaire Raisonné des Sciences, des Arts, et Métiers*. Paris: A. Breton, M-A. David, L. Durand and A-C. Briasson.

D'Alembert, J.L.R. (1759). *Essai sur les éléments de philosophie*. Paris.

Dee, J. (1570). *The Mathematicall Praeface to Elements of Geometrie of Euclid of Megara*. London: John Daye.

Dee, J. (1582). "Dee's diary for 24 November 1582", in Sherman, W.H. (1995). *John Dee: The Politics of Reading and Writing in the English Renaissance*. Amherst: University of Massachusetts Press, p. 51.

Dupuy, L. (1777). *Fragment d'un ouvrage Grec d'Anthémius, sur des Paradoxes de Mécanique*. Paris.

Evelyn, J. (1664). *A Parallel of the Antient Architecture with the Modern, from the French of Roland Fréart, to which was added an Account of Architects and Architecture*. London.

Fowler, D. (1999). *The Mathematics of Plato's Academy*. Oxford: Clarendon Press.

Galileo, G. (1914). *Dialogue Concerning Two New Sciences by Galileo Galilei*. Translated from the Italian and Latin into English by Henry Crew and Alfonso de Salvio. New York: Macmillan.

Gingerich, O. (1990). "Biography of Kepler," in *Dictionary of Scientific Biography*. New York: Charles Scribner's Sons.

Ginovart, J.L. et al. (2017). "Hooke's Chain Theory and the Construction of Catenary Arches in Spain," *International Journal of Architectural Heritage*. 11(5), pp. 703–16.

Goffman, C. (1969). "And what is your Erdös number?", *American Mathematical Monthly*, 76(7), p. 791.

Graham, R. L. (1983). *Rudiments of Ramsey Theory*. Providence, Rhode Island: AMS.

Graham, R. and Butler, S. (2015). *Rudiments of Ramsey Theory* (2nd ed). Providence, RI: AMS.

Grattan-Guinness, I. (1982). *Psychical Research: A Guide to Its History, Principles and Practices: In Celebration of 100 Years of the Society for Psychical Research*. Wellingsborough: Aquarian Press.

Grattan-Guinness, I. (2005). "The 'École Polytechnique', 1794–1850: Differences over Educational Purpose and Teaching Practice." *The American Mathematical Monthly*, 112(3), pp. 233–50.

Gray, J. (2008). "Modernism in Mathematics," in *The Oxford Handbook of the History of Mathematics*, edited by Eleanor Robson and Jackie Stedall, Oxford: OUP.

Grothendieck, A. (1985). *Récoltes et Semailles*. Montpellier: University des Sciences et Techniques du Languedoc.

Grothendieck, A. and Dieudonné, J. (1960). "Éléments de géométrie algébrique," *Publications Mathématiques de l'Institut des Hautes Études Scientifiques*, 4(1), pp. 5–214.

Guy, R.K. (2004). "E17: Permutation Sequences." *Unsolved problems in number theory* (3rd ed.) New York: Springer-Verlag, pp. 336–7.

Hales, T.C. (1999). "The Honeycomb Conjecture." http://arxiv.org/abs/math.MG/9906042. Accessed 1 July 2019.

Hales, T.C. (2001). "The Honeycomb Conjecture." *Disc. Comp. Geom.* 25, pp. 1–22.

Halsted, G.B.(1878). "Note on the First English Euclid," *American Journal of Mathematics*, 2, pp. 46–48.

Harford, T. (2017). "Is This the Most Influential Work in the History of Capitalism?". *BBC World Service, 50 Things That Made the Modern Economy*. www.bbc.co.uk/news/business-41582244. Accessed 1 July 2019.

Heath, T.L. (1908). "Euclid and the Traditions about Him", in *Volume 1 of The Thirteen Books of Euclid's Elements*, VI. 1, pp. 1–6, edited by T. L. Heath and J. L. Heiberg, Cambridge: Cambridge University Press.

Heath, T.L. (1921). *A History of Greek Mathematics*. Boston: Adamant Media Corporation.

Heath, T.L. (1921). *A History of Greek Mathematics*. Oxford: Clarendon Press.

Heath, T.L. (1956). *The Thirteen Books of Euclid's Elements*. Cambridge: Cambridge University Press.

Heiberg, T.L. (1953). *The Works of Archimedes: With a Supplement 'The Method of Archimedes'*, edited in modern notation with introduction and annotation by T.L. Heath, New York: Cambridge University Press.

Hinton, C. (1888). *The Fourth Dimension*. London: Arno.

Hobson, E.W. (1913). *Squaring the Circle: A History of the Problem*. Cambridge: Cambridge University Press.

Iliffe, R. (2012). "The Life of Isaac Newton". *Isaac Newton Guidebook*. Cambridge: The Faraday Institute, pp. 18–25.

Itard, J. and Dedron, P. (1959). *Mathématiques et Mathématieicnes*. Paris: Magnard.

Jackson, A. (2004). "Comme Appelé du Néant – As If Summoned from the Void: The Life of Alexandre Grothendieck", in *Notices of the AMS*, vol. 51(10), pp. 1196–212.

Ji, L., and Papadopoulos, A. (2015). *Sophus Lie and Felix Klein: The Erlangen Program and Its Impact in Mathematics and Physics*. Zurich: European Mathematical Society.

Kennedy, I. G. (2016). *The Booles and the Hintons*. Atrium.

Kepler, J. (1596). *Mysterium Cosmographicum*. Graz.

Kepler, J. (1619). *Harmonices Mundi*. Linz.

Kepler, J. (1937). *Johannes Kepler Gesammelte Werke*, vol. 3, 108. Munchen: C.H. Becksche Verlagsbuchhandlung.

Kepler, J. (1967). *Kepler's Somnium: The Dream, Or Posthumous Work on Lunar Astronomy*. New York: Dover Publications.

Kepler, J. (1975). *L'étrenne ou la neige sexangulaire*. Translated from Latin by R. Halleux. Paris: J. Vrin éditions du CNRS.

Knorr, W. (1983). "The Geometry of Burning-Mirrors in Antiquity", *Isis*, vol. 74, no. 1, pp. 53–73.

La Roche, E. (1520). *L'arismetique nouellement composée*. Fradin.

Lagrange, J. (1797). *Théorie des fonctions analytiques, contenant les principes du calcul différentiel*. Paris: Imprimérie de la République.

Lawrence, S. (2016) "What are we Like…," edited by B. Larvor. *Mathematical Cultures: the London Meetings 2012–2014. Trends in the History of Science*. Basel: Birkhauser.

Lawrence, S. (2015). "Life, Architecture, Mathematics, and the Fourth Dimension." *Nexus Network Journal*, 17, pp. 587–604.

Lawrence, S. (2011). "Dee and his Books," *BSHM Bulletin: Journal of the British Society for the History of Mathematics*, 26(3), pp. 160–66.

Lawrence, S. and McCartney, M. (eds.) (2015). *Mathematicians and Their Gods: Interactions between Mathematics and Religious Beliefs*. Oxford: Oxford University Press.

Lehmann, K. (1945). "The Dome of Heaven," *The Art Bulletin*, 27(1), pp. 1–27.

Lloyd, D.R. (2012). "How Old are the Platonic Solids?" *BSHM Bulletin: Journal of the British Society for the History of Mathematics*, 27:3, pp. 131–40.

Mazzotti, M. (2007). *The World of Maria Gaetana Agnesi, Mathematician of God*. Baltimore: Johns Hopkins University Press.

Möbius, A.F. (1827). *Der barycentrische Calcul*. Leipzig: Verlag von Johann Ambrosius Berth.

Moore, J.T. (1967). *Elements of Abstract Algebra, Second Edition*. Toronto: Collier-Macmillan.

Moyon, M. (2018). "Dividing a Triangle in the Middle Ages: An Example from Latin Works on Practical Geometry". In É. Barbin et al., *Let History into the Mathematics Classroom*. New York: Springer International.

Netz, R. and Noel, W. (2007). *The Archimedes Codex: How a Medieval Prayer Book is Revealing the True Genius of Antiquity's Greatest Scientist*. Cambridge, Massachusetts: Da Capo Press.

Neuenschwander, D. (2010). *Emmy Noether's Wonderful Theorem*. Baltimore: John Hopkins University Press.

Newton, I. (1995). *The Principia*. Amherst: Prometheus Books.

Noakes, R. (2004). "Spiritualism, Science and the Supernatural in Mid-Victorian Britain". *Cambridge Studies in Nineteenth Century Literature and Culture*, 42, pp. 23–43.

Noether, E. (1921). *Idealtheorie in Ringbereichen*, translated and published online in 2014, arxiv.org/pdf/1401.2577v1.pdf. Accessed 1 July 2019.

Ohm, M. (1835). *Die Reine Elementar-Mathematik*. Berlin: Jonas Veilags-Buchhandlung.

Pacioli, L. (1956). *De divina proportione*, edited by Biggiogero, Fontes bibliothecae Ambrosianae, 31, Milan.

Pacioli, L. (1994). *Particularis de Computis et Scripturis*. A Contemporary Interpretation, translated by Jeremy Cripps. Seattle: Pacioli Society.

Pindar (1997). *Olympian Odes, Phitian odes,* edited and translated by W.H. Race, Loeb Classical Library 56, Cambridge, MA: Harvard University Press.

Plato (1997). "Timaeus", *Cosmology*, translated by Francis M. Cornford, Cambridge, MA: Hackett Publishing Co.

Plato (2009). *Meno*, translated by Robin Waterfiled, *Meno and Other Dialogues*, Oxford: Oxford University Press.

Proclus, D. (1792). *The Philosophical and Mathematical Commentaries of Proclus, etc,* translated by Thomas Taylor. London.

Ramsey, F. (1928). "On a Problem of Formal Logic", *Proc. London Mathematical Society,* 30, pp. 264–86.

Sanders, P.M. (1990). *The Regular Polyhedra in Renaissance Science and Philosophy.* PhD Thesis, University of London, Warburg Institute.

Sarton, G. (1951). "When is the Term 'Golden Section' or its Equivalent in Other Languages Originate?" *Isis,* 42, p. 47.

Scattergood, J. (2015). *The Complete English Poems of John Skelton.* Liverpool: Liverpool University Press.

Schiffman, Z.S. (2011). *The Birth of the Past.* Baltimore, MD: Johns Hopkins University Press.

Shakespeare, W. (2008). *The Tempest.* London: Red Globe Press.

Simms, D.L. (1977). "Archimedes and the Burning Mirrors of Syracuse", *Technology and Culture,* 18(1), pp. 1–26.

Smith, A. (2004). *Making Mathematics Count. The Report on the Inquiry into Post-14 Mathematics Education.* London: HMSO.

Sombart, W. (1924). *Der Moderne Kapitalismus,* 6th ed., vol 2(1), pp. 118–19. Munich and Lepizig.

Stakhov, A.P. (2009). *Mathematics and Harmony From Euclid to Contemporary Mathematics.* London: World Scientific.

Stringham, W. (1880). "Regular Figures in n-Dimensional Space," *American Journal of Mathematics,* 3, 1:114. Baltimore: John Hopkins University Press.

Tent, M.B.W. (2008). *Emmy Noether: The Mother of Modern Algebra.* Wellesley, MA: A.K. Peters.

Van Oss, R.G. (1983) "D'Alembert and the Fourth Dimension", *Historia Mathematica,* vol. 10, no. 4, pp. 455–57.

Vardi, Ilan. (not known). *Archimedes, The Sand Reckoner.* École Polytechnique.

Weil, A. (1991). *The Apprenticeship of a Mathematician.* New York: Springer.

Westman, R.S., and McGuire, J.E. (eds.) (1977). *Hermeticism and the Scientific Revolution,* papers read at a Clark Library Seminar, 9 March 1974, William Andrews Clark Memorial Library.

Wittkower, R. (1960). "The Chancing Concept of Proportion", *Daedalus,* 89. pp. 199–215. Chicago: The University of Chicago Press, HSS.

Woodward, W.W. (2011). *Prospero's America, John Winthrop, Jr., Alchemy, and the Creation of New England Culture,* 1606–1676. Chapel Hill, NC: The University of North Carolina Press.

Wren, C. (1750). *Parentalia*, Tract I, p. 351.

Yates, F. (1966). *The Art of Memory*. London: Routledge and Kegan Paul.

Yates, R.C. (1952). *A Handbook on Curves and Their Properties*. Ann Arbor: J.W. Edwards.

Zöllner, F. (1878). *Uber Wirkungen in die Ferne*, Leipzig: Wissenschaftliche Abhandlungen.

Index

Note: Page numbers in *italics* refer to figures. Footnotes are denoted by n, e.g. 66n2.

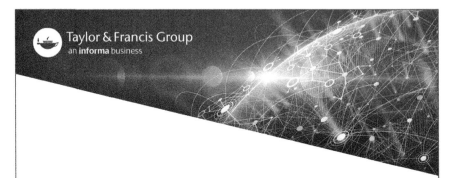

Printed and bound by CPI Group (UK) Ltd, Croydon, CR0 4YY

23/10/2024

01778262-0007